THE 10 COMMANDMENTS SOFTWARE TESTING

THE 10 COMMANDMENTS OF SOFTWARE TESTING

DONNA L. SIMMONS

TATE PUBLISHING & *Enterprises*

The Ten Commandments of Software Testing

Published by Tate Publishing & Enterprises, LLC
127 E. Trade Center Terrace | Mustang, Oklahoma 73064 USA
1.888.361.9473 | www.tatepublishing.com

Tate Publishing is committed to excellence in the publishing industry. The company reflects the philosophy established by the founders, based on Psalm 68:11,
"The Lord gave the word and great was the company of those who published it."

Cover design by Kristen Verser
Interior design by Lindsay B. Behrens

Published in the United States of America

ISBN: 978-1-61777-136-1
1. Computers / Software Development & Engineering / Quality Assurance & Testing
2. Computers / Software Development & Engineering / General
11.06.30

TABLE OF CONTENTS

INTRODUCTION

A long-time friend, Brian Whitfield of Essential Resources, and I were chatting one day, and he mentioned to me that I should write a book about my quality assurance (QA) experiences and to harvest on those twenty-five years of experiences. There are some experiences I would like to share with readers: some best practices, considerations, recommendations, successes, failures, solutions, examples, and templates for planning and executing of software testing. This book is written for all my colleagues whose desire was to deliver a software product that would meet quality standards but have suffered with being chastised or the loss of their jobs for the following reasons:

- They had the courage to speak out about not delivering a product that was not working due to lack of defect prevention

- Their immediate management had minimal QA experience and were threatened by the QA resource's knowledge and experience
- They did not meet an unreasonable deadline
- They did not have automated tool experience, and there was no budget for training to learn the tool.

In the many years of practicing quality assurance, it appears that "defect free" is a remote thought by many in information technology organizations, and for some it is an ideal that is not reachable in today's technology. However, once technology organizations are willing to accept the delivery of defects in a product, then the questions have to be asked: How many defects are they willing to deliver, or how severe of a defect are they willing to deliver? What impact to the client are businesses willing to impose on the client, or conversely, how much of a risk is the client willing to accept to their business when a product contains defects? The philosophy of offering the opportunity to deliver products that contain defects provides the opportunity for an open door to deliver inferior products that contain defects and the customer is reduced to dealing with it.

Now, if you think about this for a moment or two, would you like it if you turned on the television set and periodically a channel skipped or part of a face did not appear? Absolutely not! The television set would be returned, or maybe a recall notice would be sent out like they do for automobiles. However, have you heard of a recall notice on software? Absolutely not! Oh, sometimes in the contractual agreements there is a warranty period established where the company that is producing the software will resolve defects found during the warranty; however, after that period of time, new agreements may be drafted for a future release. What this means is that it will cost the client more money to have those defects resolved unless it was established in the contracts or in the service level agreement (SLA) when the project was negotiated.

It is difficult today to deliver defect-free software, but I have, in my career, delivered a defect-free product. What is interesting is that when I have delivered presentations on testing and on my credentials, which address the fact that I have delivered a defect-free product, not one person has ever asked the question, "How were you able to deliver a product that is defect free?" We are so used to delivering software products that contain defects or bugs that it is understood we will deliver a product with defects. To soften the blow of a defect or a bug, euphemisms are used, such as an *anomaly*, an *issue*, or a *problem*.

Certainly, in this world of globalization there is a need for rapid application development (RAD) to be able to compete with the world marketplace. However, with that in mind, information technology organizations have to work smarter than ever before by delivering the product more quickly to the client.

I have found that in most cases the corporate world has a vision to deliver quality software applications; however, there appear to be a number of indications that they really do not want to do this, as various factors come to play that steer the decision makers in changing their philosophical quality course. Certainly the cost of producing the product and their return on investment plays heavily in delivering a quality product to the customer.

Generally speaking, it has been bantered about over the years that marketing or sales organizations do not have a clue in what it takes to produce a software product. They will quickly produce a date, which in some instances had those of us in QA wondering if they did not pull a date out of their hat without any considerations for the amount of time it would take to develop the software product. Unfortunately, the information technology (IT) organizations have to abide by the delivery date the marketing or sales team proposes to the client, instead of taking the time to figure out the appropriate delivery date based on the size and scope of the project.

It is not uncommon for the client to demand an unrealistic delivery date or for marketing or sales agrees to unrealistic delivery dates, and then the information technology organization has to make decisions that result in products that are delivered with defects. As soon as the major software development processes are reduced in scope, then quality is jeopardized. The question has to be asked: When will the executive level take a stand and be willing to deliver products that are not defect ridden and take the time to have the resources to coach the customer?

Some years back, I attended a conference where one of the presentations was about eXtreme programming, and the developers that were presenting the program indicated that they did not use QA testers. Their model consisted of a developer coding and another looking on to ensure the coding met the review process and that customer's requirements were met. I recall stating in that conference session that QA was set back thirty years, and it was not long before I heard mumbling throughout the audience: "She is right; she is right."

Now with the advent of the Agile model, companies believe they have found another silver bullet. This time it appears we have given developers carte blanch to deliver products without all the quality assurance processes and procedures. Not too long ago, I attended a presentation where the firm was discussing their Agile model. Unfortunately, requirements were being drafted as the development organization was coding, as this followed their Agile process. This method has the developer work in tandem with the customer. The customer provides the requirements in the working sessions, and the developer writes the code based on the provided requirements. My concern about this methodology is whether the requirements are being fully developed to the granularity that will avoid a lot of rework. And for those organizations that are out there that the customer is not heavily involved in the Agile process, and the QA organization is not a member of the development process, then quality of the software product that is being developed is impacted. Alas, we have given developers exactly what they have totally desired: no independent QA organization and they have total control over the code.

I fully understand the need to meet delivery dates and to find methods that will implement software more quickly. However, the cost of quality is constantly being affected with the need for rapid development when requirements are not fully elicited. According to the Quality Assurance Institute's Certified Quality Assurance Common Body Knowledge (CBOK):

> The cost of quality (COQ) is the money spent above and beyond expected production costs (labor, materials, equipment) to ensure that the product the customer receives is a quality (defect-free) product. NOTE: This includes the cost of repairing a defective product that was shipped to a customer and the associated damage costs.[1]

Certainly, with the current economic crisis, many best practices may be in jeopardy for a number of reasons. We are now witnessing and feeling the impact of improper planning. Businesses are looking at every possible means to increase their profit margin. With the needs of a world-wide marketplace, businesses are under the gun to produce more quickly. With this consideration, the business sector must embrace principles of integrity in the delivery of software applications that are not riddled with defects.

"There is a true glory and a true honor, the glory of duty done, the honor of integrity and principle."

Robert E. Lee

THE PURPOSE AND BENEFITS FOR GOOD TESTING!

The purpose of testing is to measure the quality of an application and its logic, and this is achieved by predetermined criteria and specifications. Testing is an activity to prove that a software application, system, or configuration is *not* working, not that it does work.

Testing is about:

- Exercising the software application, system, or program
- Establishing confidence that a program does what it is supposed to and doesn't do what it isn't supposed to
- Analyzing a program with the intent of finding defects
- Measuring system functionality and quality
- Evaluating the attributes and capabilities of programs, project work products, and assess-

ing whether they achieve required or acceptable results,

- Performing inspections and structured peer reviews of requirements and design, as well as execution of the application's code.[2]

Why should we worry about good testing of an application? According to the developers, they have designed and built the application to the customer's requirements. Doesn't the definition of *requirements* mean meeting or exceeding the customer's needs or expectations? Well, that is not always the case, as it is not uncommon the designed product does not represent what the customer has defined. There have been times during testing that a feature or function has been added by the developer that was not documented in the functional or design requirements; it was something the developer thought would be nice to include in the software. Of course, this would be a defect, right? Correct, as the nice-to-have feature or function was not requested by the customer. Then it would be the responsibility of the project or product manager to discuss this with the customer as to whether it should remain in the software. If this is the case, the business believes the respective change shall be in the application, and then the change must follow the change control process.

In my earlier years of working in information technology, I held a dual role as a business analyst and as a QA tester, and it was not uncommon to be asked to demonstrate the benefits of the features and functions of the application to a future customer's business needs. The intent of the demonstration was to prove to the customer that the software worked and it would still work with their requested modifications.

As a software tester, whether it was testing applications on the mainframe, mid-range (AS400), client, or web environment, measuring the quality of the application was the primary focus. With defect detection as the primary focus, then, by definition, testing is quality control. Quality control (QC) is the process by which product quality is compared with applicable standards and the action taken when nonconformance is detected. Its focus is defect detection and removal. However, it is a very common thought that quality assurance means testing or defect detection, whereas the actual meaning of quality assurance is a set of support activities needed to provide adequate confidence that processes are established and continuously improved in order to produce products that meet specifications and are fit for use. These activities include facilitation, training, measurement, and analysis; however, nowhere in this definition does it imply that defect detection is QA. Defect detection is purely quality control. The question needs to be asked, why aren't the software test teams called QC teams instead of QA teams?

When I was studying for my Certified Software Quality Analyst certification, there was a chapter in the *Certified Body of Knowledge* (CBOK) that clearly outlined the differences between QA and QC. When the activity is defect driven, then we are discussing quality control; quality assurance is a set of activities. The differentiating of quality control with quality assurance is outlined in Commandment Ten: There Shall Be Defect Management.

Defect prevention really depends upon the leadership of the organization. When the leadership of an organization fully understands the importance of defect prevention, preventative methods are established to ensure that quality occurs at the beginning of the software development cycle, not just during the testing phase of the project. When an organization implements defect prevention processes, the cost of quality is reduced, and benefits from the processes. *Figure 0.1* highlights the purpose and benefits of testing over a number of decades.

Purpose and Benefits of Good Testing			
DEMONSTRATION **Show it works** **1960s**	**DETECTION** **Find Defects** **1970s**	**PREVENTION** **Manage Quality** **1990s**	**COMBINED QUALITY METHODS** **2000s**
Gain Confidence the system can be used with acceptable risk Try out features and functions under unusual conditions and situations Assure a work product is complete and ready for use or integration .	Discover defects and system deficiencies Define system capabilities and limitations Provide information on the quality of components, systems and work products	Clarify system specifications and performance Provide information that prevents or reduces the likelihood of errors being made Detect errors earlier in the process Identify risks and problems and ways to avoid them in the future[i1]	Includes all the activities from the 70s, 80s and 90s Manages test efforts that are responsive to today's global business needs Utilizes tools, techniques and software methodologies that are more efficient and effective for testing Rapid Application Development (RAD) has become more of a focus

We can easily see that the process has evolved from demonstrating that the software works to measures that will prevent defects. However, with the advent of the year 2000 and the testing that needed to occur, it was believed that IT organizations would have learned the importance of quality assurance in this massive effort to modify all the code that needed the date logic renovated, and the processes and testing procedures that were put in place would certainly continue. Wrong! Many of my colleagues believed that IT organizations fully understood the importance of good testing. Well, not in the years following 2000. The businesses were now trying to find methods that would enable IT organizations to implement more quickly and, if possible, reduce QA efforts that include eliminating QA activities and rely on the developers for testing.

Today companies are looking at more efficient and effective methods for software testing, and rapid application development (RAD) is being used to get it out quicker and faster to the customer. In some instances, the customer performs testing to validate what was released does work or logs the defects and reports the bugs to the project team to resolve. One of the models that is being more extensively used is the Agile model, and this model depends on the users to be involved with the requirements elicitation and validation of the software. Some projects have a QA resource assigned when the code has been constructed for one of the "sprints" (a period of time anywhere from two weeks to four weeks where the development team delivers agreed upon portion of the software). However, it appears in a number of companies they are depending on the developer to do what is right in testing their software. I guess a person could equate that to placing a "fox in the hen house."

What other silver bullet could be employed to demonstrate the benefits of good testing? Many organizations have purchased automated tools to create test scripts that will test the software application. Certainly, the tests will be repeatable, and the probability for errors being introduced by a manual tester can be reduced when a manual test case is not replicated exactly as scripted during test execution. Therefore, automated test tools must be the answer to ensure the product works. Wouldn't you think this to be true? The answer is it depends. It depends on the:

- Selection of the right tool for test execution
- Test environment
- Type of tests being executed
- What needs to be tested
- The skill sets of the tester drafting the test scripts
- The amount of time for building automated test scripts

I am not trying to say that automated tools do not work; that is not true. What I am trying to state is that automated tools have their place in the testing effort but are not the only solution.

Conceptually, the benefit of using an automated tool is supposed to reduce the amount of time for test preparation. As a cautionary statement, automated test tools are not necessarily the silver bullet for reducing the amount of time it takes to test, as sometimes this can take longer than what it would take to code the application. If you are planning to use automated tools for testing your code, then the timeframe has to be seriously addressed based on the number of requirements that have to be tested, the number of automated software test engineers that are available to create and execute the test scripts, and the knowledge and skill level of the test engineer in the automated tool of choice. All of these considerations may have an impact on the timeframe for executing automated testing of a software application.

In many cases, the automated test tools are effective in reducing the amount of time for regression testing, as to re-execute all those manual tests for regression and the number of regressions can have a negative impact on the timeline. A cautionary statement needs to be addressed: if the test engineer is only executing record and playback, that is not the best use of an automated tool that can benefit the test organization. The automated test tools have a much greater capacity for testing the functionality of a system, measuring the back end of the application and its performance capabilities. The performance tools create graphical reports during the test procedures that provide the metrics for tuning servers, mainframes, midrange computers, and databases for the IT organization. In this instance, this is good testing if the test engineers that execute performance testing have the experience and knowledge to leverage the results the test tool is generating to tune the servers and hardware platforms.

How can we know that our test will be a good test? The answer is proper planning, processes, and procedures. The software tester must be assigned when the business case is approved and be "joined at the hip," so to speak, with the business analyst and the development team to garner all the information that is needed to develop the test plan and test cases. This partnering with the business team and the development team is exceptionally important, as it builds a relationship that can open up more communication with the team. If the tester is not included in all the discussions and meetings, important facts may not be considered for test and may not be addressed. The end result is the client may receive a product that does not perform as required.

On a project that I was assigned, the project manager would not allow QA involvement at any of the project meetings. Since there was no requirements document drafted, I was to use the prototype that was deemed to be representative of the customer's requirements for the project. To draft the test plan and test cases, I worked with the business analyst (BA) and the human factors resources that were assigned to the project. Communication with the rest of the development team was very limited and definitely created some division with the QA organization. In order for the project to be a success, all parties must be part of the team and communicate with one another on a regular basis to avoid gaps with the project. *Figure 0.2* is a pyramid test preparation model that identifies at a very high level what is needed for test preparation.

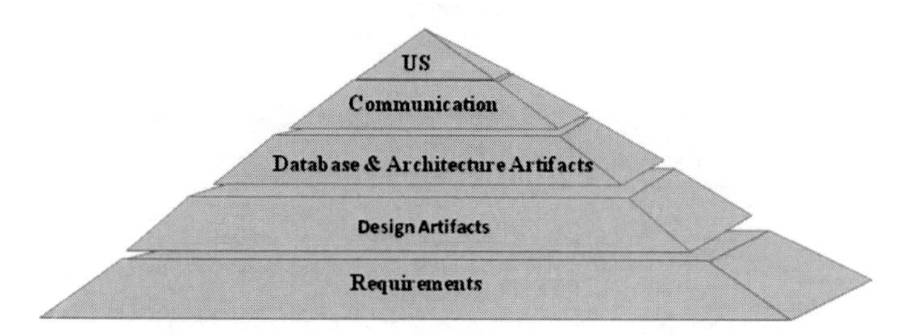

Figure 0.2–A High Level Test Preparation Model

As many of us know in the information technology industry, requirements are the building blocks for the development of a software application. All development and testing is totally dependent upon a rigorous requirement elicitation process in order for the requirements to be completely defined, reviewed, and then approved in order for the software product be delivered successfully. It is imperative that the quality assurance (QA) analyst be assigned as soon as the requirements are developed, or in the best case scenario, be part of the requirements elicitation process. When the QA tester is assigned at that point in time of the software development life cycle (SDLC), the QA analyst has a complete understanding of the requirements and then is able to leverage that knowledge for planning and authoring of test scenarios and or test conditions and test cases to test the software application.

In preparation of the design artifacts, database, and architectural artifacts, the QA tester must be invited to the design meetings and be provided with a copy of all the design artifacts in order to plan for the test and to create the test deliverables that will validate the approved design requirements.

In my history of performing testing activities, I was assigned to a project where the development side of the project team did everything in their power to avoid having the QA resource as a partner in the development process. This project included the development of a new product and the conversion of data. Throughout the development process I indicated we would need to perform data validation of the converted data. Of course, the pat answer was that we did not need to handle that because development would be involved with that activity.

I continued the plan of validating the data, and when the project was finally turned over to QA, the data was erroneous, as was suspected. There were duplicate rows of data, truncated data, missing data, and invalid data. Needless to say, if the development team would have opened up the lines of communication with QA and other IT departments, the project would not have been in the troubled state it was in when system testing began.

In essence, this was a communication issue, as the project planning and managing did not properly communicate the needs for the test phase of the project, and allowed the development staff to handle it. A communication plan should have been drafted that would have aided all members of the team and other departments that were involved with the project

To develop a software application, it takes a marriage between many groups of resources to work effectively and efficiently. This requires the building of a working partnership to ensure the project moves forward successfully and to deliver a quality product. It takes all of "us" that are part of the team.

"Let us never negotiate out of fear, but let us never fear to negotiate."

John F. Kennedy

COMMANDMENT ONE:

Tests That Validate Requirements Must Be Traced to Approved Business and Design Requirements and Be Executed by an Independent Group

This commandment contains four very important statements. The first part of the tenet addresses the need for tests that validate requirements. Basically, there are two methods of testing the requirements, and this is achieved by performing requirements reviews, as this is the first point of insertion where a defect can be inserted into a business artifact.

The second part of the statement addresses the fact that the tests must be traced to the approved business requirements and design requirements. So let us first address the term *approved.* In this case I am stating the business and design requirements must be approved by the business stakeholder. If approval and signoff does not occur, then the project will not know if the requirements meet the businesses needs and expectations. The review and approval process ensures the appropriate resources are reviewing the requirement deliverables that meet the business and technical needs.

In the requirements phase of the project, is the first point of insertion a defect is introduced into the development of the software product. At this phase of the project, the fixing of those defects is the least costly in comparison to when the defect is encountered in the system integration test phase or, in a worst-case scenario, in production. The cost of quality (the cost of development plus rework) goes up exponentially when the defect is found in production in contrast to when it is discovered during the requirements phase of the project. The phases for software development are Conception, Requirements, Design, Construction, Test and Installation.

What happens when requirements are not reviewed and approved by the business stakeholder? *Figure 1.1 Requirement–Swing* provides a graphical image of what various departmental resources perceive is being requested by the customer. Obviously, the requirements were not reviewed in a formal review process. We easily can see that if the QA analyst would have created the test cases that were based on any of the images with the exception of the tree with the tire swing, a product would have been delivered to the customer that did not meet the customer's requirements and expectations.

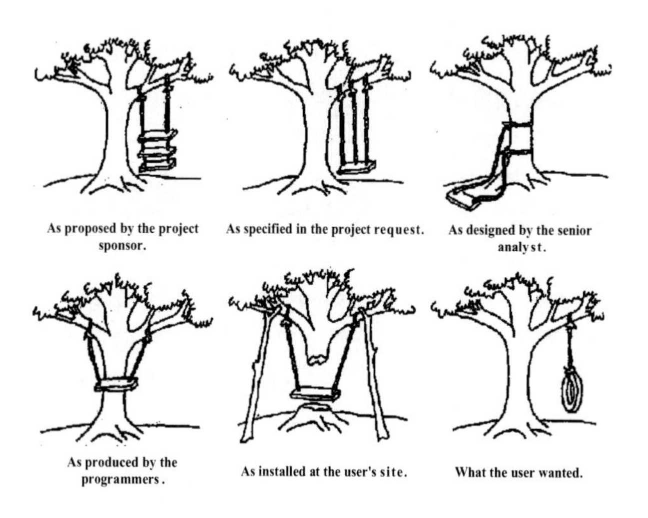

Figure 1.1 Tire Swing Cartoon

The development of a test plan and test cases is not a simple task, and it requires planning by the QA analyst. In some organizations, the task of authoring a test plan is the responsibility of the QA test lead. The authoring of the plan requires the capability to communicate and work effectively with all the departments that would provide the needed information for the test plan. This is a time-intensive task and requires involvement with resources from the business and development to provide the necessary information so that the QA analyst can author the test plan. A subsequent task would be the creation of test cases or test scripts, and based on the technology, scope, and the various types of tests, this will require various skill sets to draft the appropriate test cases.

The third part of the commandment is tracing the test cases to the approved requirements. Traceability does not begin with the test phase; it begins when the requirements are approved. The test conditions and/or test scenarios are then identified, and the planned tests are then traced to the business and design requirements. For the sake of understanding, it is the author's position that test cases are manually created and test scripts are created through the use of an automated tool.

How are the test cases traced to the business and design requirements? Traceability may occur with the use of an automated tool, and this is the best method to employ, as the manual process is time-intensive to handle this activity with an Excel spreadsheet. The likelihood for gaps to increase during the traceability activity of tracing the requirements to other artifacts, such as use cases, test cases and other design artifacts. There are many traceability products that are available for a business to evaluate and determine which would be the best solution for their organization. In order for a business to provide an automated tool, the organization needs to address its business needs and determine which tool would be in their best interest. I have found not all companies' budgets afford the opportunity to purchase these tools, and it is not just the cost of the software; it also includes the cost of additional licenses for all the resources that will be using the tool, maintenance agreements with the software company, and the cost of training the resources on the tool of choice.

Who will be using the requirements traceability tool, and how many of the staff will need training on the chosen software tool? Who will perform the training? Will it be in-house or on location where the training programs are held? Then travel expenses will also have to be considered as part of the cost of purchase. The concept of train the trainer does not always work. Train the trainer is employed in many companies, and this occurs when one person is sent to a class and is to train the rest of the staff on what was learned. So often the knowledge stays with the trainer and is not disseminated to the rest of the team.

For companies or IT organizations to get the "best bang for the buck" is to hire the resource(s) that already know how to use the tool, and part of the hiring agreement is they will be required to train the other resources that will be using the tool. There is a piece of caution here: don't just hire the candidate that says he or she has knowledge, but pose questions that will challenge the candidate's skill sets, knowledge, and practical experience with the automated tool of choice. It may cost more money to get the resources, but in the long run it will be more effective and less costly.

Then there is open source software (also known as freeware) on the Web that can be downloaded for building a traceability matrix. However, in the circumstance when your company does not allow the downloading of any open source software, then the other solution is to create a manual traceability matrix. Tracing manually is generally handled with an Excel spreadsheet, and this can be a cumbersome process, but it is also an effective method to use. *Figure 1.2a and Figure 1.2b Excel Traceability Matrix* provide an example of tracing project artifacts.

<Project Name – Project Number>					
Requirements Number	**Requirements Short Description**	**Functional Specification Number**	**Functional Specification Short Description**	**System Requirements Specification Number**	**System Requirements Short Description**
100	Contact Customer Service	100.11	Contact Us Hours of Availability	100.21	Contact Us Available Hours of Operation
100	Contact Customer Service	100.11	Contact Us Hours of Availability	100.21	Contact Us Available Hours of Operation

<Project Name – Project Number>						
Use Case Number	**Use Case Short Description**	**Business Rule Number**	**Business Rule Short Description**	**Program/ Module Name**	**Test Case Number**	**Test Case Short Description**
100.31	Contact us 24/7 by telephone	100.41	Hours of Operation telephone contact 24/7	CC123456	100.51	Validate contact customer care 24/7 by telephone number 800.999.9999
100.31	Contact us 24/7 by email	100.41	Hours of Operation email contact 24/7	CC123456	100.51	Validate contact customer care 24/7 by email at email address xxx@xxxxx.com

Figure 1.2a & 1.2b Traceability Matrix

Before you begin creating the matrix, some planning needs to occur as to how you will structure the spreadsheet. All the artifacts that are used for the development of the software are to be evaluated for the traceability activity.

So often it is perceived as the responsibility of the business analyst and the QA analyst to trace the business, design, and test artifacts. Well, in truth, it is the responsibility of all the departments. This includes business analysts, designers, architects, developers, and software test analysts/software QA analysts—whoever has had a part in the development of the application's artifacts. Traceability is to be provided for all artifacts that are involved in the development or modification of the software application. It is the responsibility of the developer, the designer, architect, and the QA analyst. If an asset is being developed that defines how the product is to be designed, developed, and tested, every resource in the workflow has the responsibility to document their respective artifacts in the traceability matrix.

The question may be asked: Why doesn't the developer create unit test cases? Some organizations do require unit test cases; however, in most cases they are not required as long as the developer does test their code. However, to ensure the code has been tested, I have requested the development organization to create a test matrix that would identify all the tests the developer was performing. The unit test matrix is then stored in the configuration management tool as a corporate asset. The traceability matrix is updated with the components that have been unit tested.

The fourth part of the commandment addresses the need for a testing team to function as an independent group. This means the test organization does not report to the development team or to the project manager.

When the QA team reports to the development manager or to the project manager their objectivity can be influenced. It is not uncommon for the project manager or the development manager to reduce the testing timeline because it has taken the construction phase longer than originally planned.

The project manager's focus is to meet the date and to deliver the product on time. I have to admit this is not an unreasonable focus; however, the question that I have had is how much value the project manager puts in testing when:

- A test type is not executed because they do not see a good reason to execute that test type
- The size and scope of a regression is reduced
- The severity level of a defect is changed to a severity level with lesser impact on the project
- The timeline is reduced for the test phase and this includes the time for test planning, test case creation and test execution
- An application is installed even though there are defects that are not resolved, and will have a negative impact on the functionality of the application

When the testers report to the project manager, they are not functioning as a separate team; the tester is then a member of the development organization and is not functioning as an independent group. Think about this for a moment: Would the development organization consider developing a product without a development manager to support them throughout the project? Would operations not have an operations manager throughout the project? The answer, of course, is most likely not.

If a tester reports to the project manager and the project manager expresses the desire to change the amount of time allocated for testing or to deliver a product with defects, there is no one to support the tester. The test team must have a manager who is willing to enter into dialogue with the project management team and the development team on behalf of the test team.

Of course, the ideal QA structure would be an executive of quality assurance (this position would report to the chief officer of operations), a director of quality assurance, a quality assurance manager to manage the software quality assurance (SQA) activities (the team's role is to perform SQA activities, such as adding measurements to the process improvement activities), and a quality assurance test manager (the test team's role is to validate the business and design requirements). When this type of structure is in place, then the test organization is an independent organization.

It is still common practice for IT organization to believe that the developer can be objective enough to perform the testing of the application he or she is designing and coding. Now, I would like you to think about this for a moment. It is very seldom that a developer will code the whole application. Generally, he or she is responsible for developing a small section of the code. Until all those pieces are integrated together, the developer would not know if the piece he or she coded would work with all the components of the application. The developer shall be responsible for unit and integrated testing. Certainly, there are other types of tests that need to be performed prior to turning over the application to the test organization. The development team may have to execute some benchmark performance tests on the servers or hardware platforms, as well as other tests that depend on the technology. When the development team has completed their testing, then the software application is turned over to the independent test group. If you would think about it from this perspective, the development team is responsible for a piece of the pie, whereas the test analyst or QA analyst is responsible for the whole pie.

Today, it is becoming more and more common for the independent group to be a group of offshore testers. In the process of considering an offshore team, it is imperative the team can function well within your organization and perform all the duties successfully that are required for the project. Some of the issues that I have encountered with offshore teams are:

- Language differences and misuse of the English language and grammar problems in writing of artifacts
- Terminology that has different meanings based on their textbook understanding. For example, I was leading a project and the term by the offshore team they used was

intimate with me, which meant they were going to communicate with me; however, I had to explain to them that the general understanding of *intimate* meant something very personal in the United States and needed to use the communication term.

- Cultural differences in how they deal with reporting issues to management
- The offshore team is not performing at its CMM/CMMI assessment level.

As a QA manager for a worldwide company that used offshore test teams for all their projects, I was requested to review a test plan and test cases the offshore test team had developed. For both products, I failed the test plan and the respective test cases. The test plan was not fully developed and did not provide the depth the project required for a system test plan. The test cases' expected results contained condition statements. A condition statement belongs in the test steps, not in the expected results. After reviewing the updated test cases a number of times, I finally decided the only way to resolve this problem was to provide a training session for an offshore company that was assessed at a CMMI maturity Level 5.

To provide the reader a brief and very high-level explanation, the Capability Maturity Model Integrated (CMMI) is an integrated framework for maturity models and associated products that integrates the two key disciplines that are inseparable in a systems development activity: software engineering and systems engineering. It is a common-sense application of process management and quality improvement concepts to product development, maintenance, and acquisition. It is a set of best practices, and serves as a model for organizational improvement.

Levels		Process Areas
5	Optimizing	Organization innovation and deployment Causal analysis and resolution
4	Quantitatively Managed	Organizational process performance Quantitative project management
3	Defined	Requirements development Technical solution Product integration Verification Validation Organizational process focus Organizational process definition Organizational training Integrated project management Risk management Decision analysis and resolution Integrated Supplier Management Integrated Teaming
2	Managed	Requirements management Project planning Project monitoring and control Configuration Management Supplier agreement management Measurement and analysis Product & Process Quality Assurance
1	Initial	

Figure 1.3 Capability Maturity Model Integrated

In the inspection process of the deliverables, in my opinion, it was determined that the offshore test team was not performing at a Level V, and to achieve Level V, the team would have had to progress through the assessment process for all the previous levels of CMMI. Obviously, their review process and training methods were flawed in reviewing test artifacts.

COMMANDMENT TWO:

The Test Process Must Be Organized in Order for Constituent Software Development Phases to Function Cooperatively and with a Defined Structure

In order for a building to be constructed, the customer consults with an architect. Based on the meeting the architect has with the customer, the architect begins to draft some requirements on what the purpose of the building is, how it will be used, the size of the building, and how many rooms it will have—in other words, the architect is beginning the process of requirements elicitation. The architect specifically needs to know what the customer wants or needs before any blueprints can be drafted. In the planning of the building, the architect must make sure that the structure integrates well with the overall plan.

Similarly, the development of an application must follow a consistent framework for moving forward in software development. You would think by this time in the world of computerization and the need to meet quality objectives, one of the first activities in planning would be to create the governance that is needed to meet the company's quality objectives when a new software application is created. Unfortunately, it is not always true that a company would establish governance for their quality objectives.

I was brought in as a consultant to perform a gap analysis on an acceptance test processes and procedures and to offer recommendations based on the findings for an online financial services business. In reviewing the company's quality methodology and the templates that were required for the acceptance test, there was no governance documentation for the business. Upon discovering there was no governance, I discussed this with the corporate quality organization, and their response that this was a new business; it was in the planning and would be forthcoming, as they were establishing a quality organization that would oversee the corporate quality objectives.

The problem was the first release of the new software did not have the governance that would have included all the corporate assets that would have lent to a more successful test of the application. The first release's test cases were not stored in the corporate repository for reuse or to build upon for system and acceptance tests for the second release. Even with the second release of the product, governance had not been established. However, plans were in place for governance to be established for release 3.0.

This framework should have been established before the first release of the project, and this should have included the policies, procedures, standards, plans, and templates for the software development life cycle. When the appropriate governance is established, it becomes a guiding light to ensure there will be a consistent structure to support the software development activities.

The software development life cycle begins with the conception phase and ends with the retirement phase. The Software Development Life Cycle phases are:

- Conception Phase
- Requirements Phase
- Design Phase
- Construction Phase
- Test Phase
- Installation and Checkout Phase
- Operations and Maintenance Phase
- Retirement Phase

Before we get into the understandings of what type of life cycle models are out there, I wish to share a thought with you about the importance of understanding what the models mean to the average tester.

When I began my QA career, I was completely green and did not even know what QA meant or what skills it took to perform QA. I was hired by a software vendor in Columbia, South Carolina, because of my insurance expertise and the ability to apply that knowledge to testing of the personal lines rating application for the independent insurance agency system. There was no known methodology and/or governance that I was aware of to support the project that was disseminated to the information technology team, also known as the "Bundle" team. The team consisted of a group of business analysts that were defined as specification writers, developers, network staff, a QA team, and an acceptance test team.

After development had completed constructing the code based on the defined requirements, the code was stored on the Novell network, which was then accessed via keying a very long path to access the logic. After the acceptance test deemed the product was ready for the customer, a gold disk was created and delivered to the customer. The process was simple and did not have a lot of governance established for the project team's objectives; however, it did work, and the products that were provided to the customer were eventually delivered defect-free.

I believe the success of defect-free products distributed to the customer was due to:

- The experience and knowledge of the business analysts to draft the requirements to the granularity that was needed for the customer
- The desire of the developers in writing code and unit testing their logic before the QA teams validated the code

- The experience and knowledge of the QA teams in performing system and acceptance tests

The Bundle Team had a tremendous amount of initiative and desire to deliver a product that was defect-free, as this was a research and development (R&D) project and all the team members had a desire for their efforts to be a successful endeavor that would improve the bottom line of the company. It was the hope of the Bundle Team they would become a permanent part of the business offerings. Unfortunately, the Bundle project did not continue, as the marketing and sales teams were not structured to market micro-computer (personal computer) products, as the corporation's focus were mainframe products.

The Waterfall model was the model of choice for most of the companies that I worked for in the beginning of my career. However, new models began to come to the forefront. Models such as Iterative and Evolutionary were beginning to be discussed as solutions for software development. I only had a cursory understanding of these models. At that time, the Internet was not there to easily understand the structure of these models and when they should be used. To make it easier for my reader, I am including some very simple understandings about these various models. In some models the information is a little more comprehensive and in others it is limited. If you find them to be of interest, I suggest finding other articles so that you can be totally versed in each of them and provide input into the discussions when a model is being considered for a project.

One of the most common models is the Waterfall model, as it is a straight-line development process, and when one phase is completed the next phase begins. There are a lot of pros and cons for the Waterfall model, and I have been on projects where the Waterfall model has been successful. The business value is delivered all at once, and the users are able to use the product in its entirety. It is like ordering a room full of furniture and all the furniture is delivered to the buyer at one time. In comparison to other iterative models, the furniture is delivered by piece or in groups based on a defined delivery schedule. However, the Waterfall Model is a time-consuming method, and unless a project has the time and funding to follow the Waterfall model, it can be viewed as too timely and costly to use for implementation.

The Waterfall model does provide the framework for processes that need to be included throughout the software development life cycle (SDLC). The Waterfall model, according to the IEEE Standard Glossary of Software Engineering Terminology, is a software development process in which the constituent activities typically are a concept phase, requirements phase, design phase, implementation phase, test phase, and installation and checkout phase, and are performed in that order, possibly with overlap but with little or no iteration.

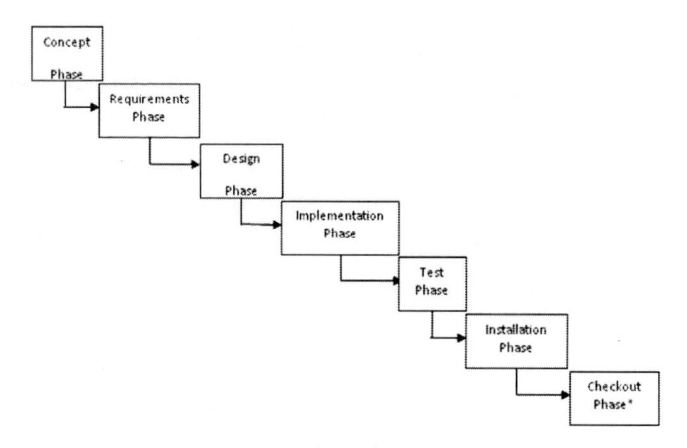

Figure 2.1 Waterfall Model

**The Checkout Phase* is also known as *Acceptance Test Phase* where the users have an opportunity to execute their test cases. The test cases in this phase are not drafted the same as they would be in a system test, which are more functionality driven, whereas the acceptance test phase the test case are more business process oriented.

Some months ago, I attended an Atlanta Quality Assurance meeting where a number of information technology executives spoke in a forum about their position on software development models, and all indicated they would not any longer walk the death walk and support the Waterfall model due to the unsure status of a successful delivery of the product to the customer. All five of them supported an iterative model as the method for developing software for the customer.

The *Iterative Model* is a model in which small iterations are conducted one after the other throughout the development cycle rather than at the start. Also, it relies on intense interaction with the business users in the prototype refinements. Often these projects will involve multiple prototyping teams working in parallel. Iterative models also rely on intense interaction with the business users in prototype refinement.

The Iterative approach is effective when:

- The client must have the application as soon as possible. The rapid iterative development is faster because the design and construction phases proceed concurrently rather than one at a time
- The client/business environment is changing rapidly. Requirements are not fully known until the end. Multiple iterations make it easier to introduce requirements changes rather than development being started with fixed requirements
- The user interface needs experimentation. A new type of application may need to experiment with several approaches to optimize the effectiveness of the user interface. An iterative approach manages this experimentation and use of client expertise to improve initial attempts.

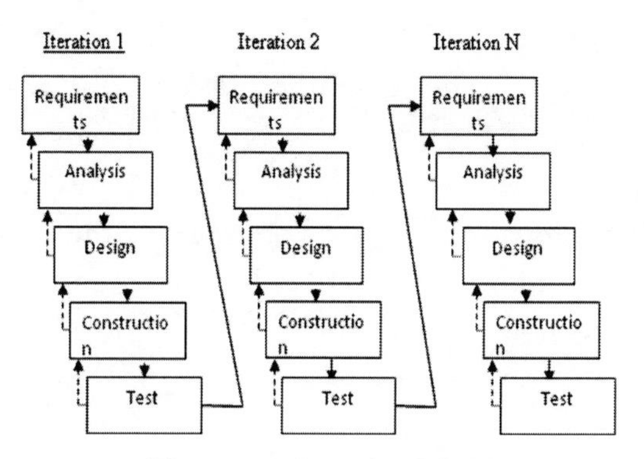

Figure 2.2 Iterative Model

The Incremental Model is used:

- To reduce the risk of project failure due to the size becoming difficult to manage
- To quickly deliver a core piece of functionality that is needed in a short time by the business without forcing the business to wait for the entire set of functions to be developed
- To allow for a proof of concept delivery of a piece of functionality that will allow a decision on committing to a much larger project
- To allow for a multi-site implementation when the sites are installed serially.

Incremental iterations are used when the full requirements are known, and there is a decision to allocate the requirements to more than one release. As each "Iteration" is completed, the project team must determine how soon "Iteration n+1" may begin based on resource constraints.

In a team where requirements, design, and construction are each being performed by different groups or individuals, the iterations can occur as described above. However, when a team consists of the same groups or individuals performing two or three of the functions, it may be more prudent to delay "Iteration n+1" until "Iteration n" is installed in Production.

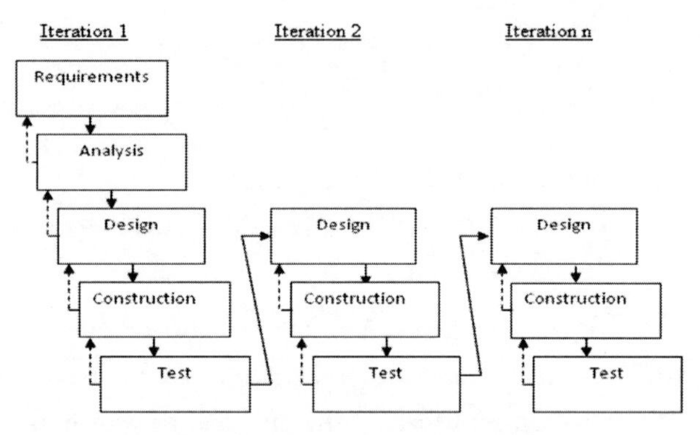

Figure 2.3 Incremental Model

The Evolutionary Model is used:

- When the requirements cannot be fully known
- When the requirements are not fully defined until the users have the opportunity to use the software. When the users have determined what other functionality is required.

In Evolutionary iterations:

- "Iteration n" the software is completed
- The Iteration is installed in Production (or a Pilot Environment)
- The users are given an opportunity to work with the software in order to determine additional requirements, prior to "Iteration n+1" beginning

The Evolutionary Model is in contrast to an Iterative-Incremental approach where the requirements are known in advance; however, a decision has been made to allocate the requirements to two or more releases. This model requires strong management control and is considered a high risk to use in software development.

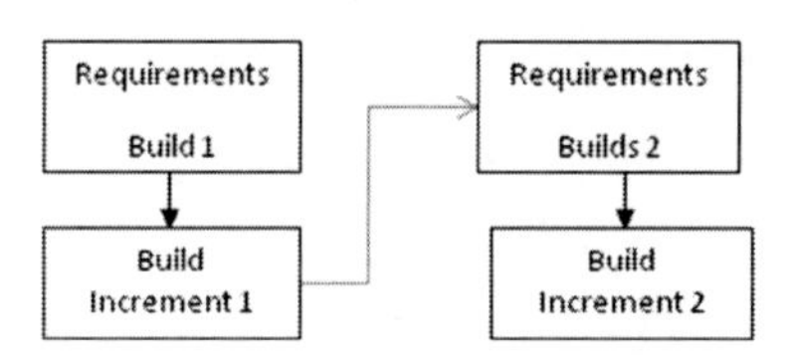

2.4 Evolutionary Model

The Spiral Model was defined by Barry Boehm in his 1988 article, "A Spiral Model of Software Development and Enhancement."

> The model is an iterative process that is performed spirally, and is generally used for long-term projects. The model is a combination of design and prototyping stages or phases. The Spiral model

> is a software development process combining elements of both design and prototyping-in-stages, in an effort to combine advantages of top-down and bottom-up concepts. Also known as the Spiral Lifecycle model (or Spiral development), it is a systems development method (SDM) used in information technology (IT). This model of development combines the features of the Prototyping model and the Waterfall model.[6]

The Spiral Model is intended for large, expensive, and complicated projects. The steps in the Spiral Model iteration can be generalized as follows:

- The new system requirements are defined in as much detail as possible. This usually involves interviewing a number of users representing all the external or internal users and other aspects of the existing system.
- A preliminary design is created for the new system. This phase is the most important part of Spiral model. In this phase all possible (and available) alternatives, which can help in developing a cost effective project, are analyzed and strategies are decided to use them. This phase has been added specially in order to identify and resolve all the possible risks in the project development. If risks indicate any kind of uncertainty in requirements, prototyping may be used to proceed with the available data and find out

possible solutions in order to deal with the potential changes in the requirements.

- A first prototype of the new system is constructed from the preliminary design. This is usually a scaled-down system and represents an approximation of the characteristics of the final product.
- A second prototype is evolved by a fourfold procedure:
- Evaluating the first prototype in terms of its strengths, weaknesses, and risks
- Defining the requirements of the second prototype
- Planning and designing the second prototype
- Constructing and testing the second prototype

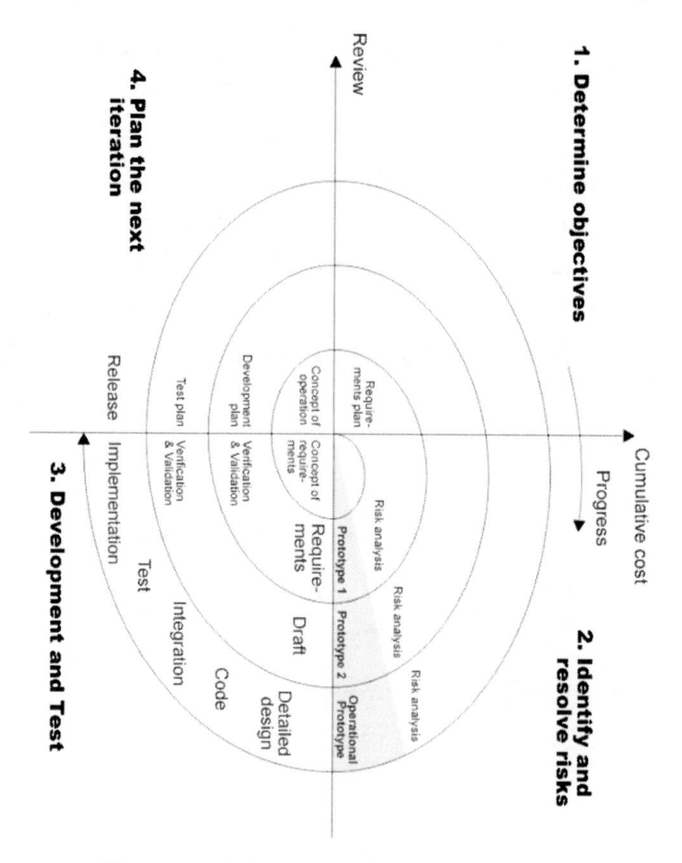

Figure 2.5 Spiral model (Barry Boehm, 1988[7]

Agile Model is a very popular model for many software organizations, as the iterations are delivered in four to six week sprints. A sprint is a short period of time that the software is defined, developed, and tested. Agile software development refers to a group of software development methodologies that are based on iterative development, where requirements and solutions evolve through collaboration between self-organizing, cross-functional teams. The term was coined in the year 2001 when the Agile Manifesto was formulated.

Agile methods generally promote a disciplined project management process that encourages frequent inspection and adaptation; a leadership philosophy that encourages teamwork, self-organization, and accountability; a set of engineering best practices that allow for rapid delivery of high-quality software; and a business approach that aligns development with customer needs and company goals. Conceptual foundations of this framework are found in modern approaches to operations management and analysis, such as lean manufacturing, soft systems methodology, speech act theory (network of conversations approach), and Six Sigma.[8]

Whichever model is used, there are quality assurance activities that must be prepared to support the quality efforts of the project for software quality assurance and for software testing. The QA department shall create a quality plan for the quality efforts for the organization and a quality assurance test plan for the type of tests that will be executed for:

- System Testing: Testing conducted on a complete, integrated system to evaluate the system's compliance with its specified requirements.[9] Note, this type of testing is also known as Integration or System Integration Test.
- Performance Testing: Testing conducted to evaluate the compliance of a system or

component with specified performance requirements.[10]

- Acceptance Testing: (1) Formal testing conducted to determine whether or not a system satisfies its acceptance criteria and to enable the customer to determine whether or not to accept the system. (2) Formal testing conducted to enable a user, customer, or other authorized entity to determine whether to accept a system or component.[11]

I have spoken with many of my colleagues about the methodology they used for software development, and in a number of companies there was no methodology in place. In these companies, there is constant rework because the requirements were not developed to the granularity that was needed for the development team to build the product based on the customer's needs. The product may end up being designed that does not represent what the customer wanted, as the developer is designing on what he or she thinks they want instead of what the customer really wanted. (See Figure 1.1 Tire Tree) The end result is the cost of quality is seriously impeded, as there are no repeatable processes in place, and there is constant rework occurring because of the number of defects that are found during test.

Or, what I believe is even worse, is when they have a posted methodology and it is mostly ignored. What I am referring to here is that a business analyst may draft the requirements, but the requirements are authored at a high level and with no granularity. The result is, in many cases, that the requirements are not testable. The development team does not perform unit testing, and the code is "thrown over the wall" to QA with the expectation that QA performs some of the unit test responsibility. In this situation, many defects are reported, and the amount of rework is above and beyond what it should be for a software project.

I recall a project I was assigned on where the flight operations portfolio had developed a software development methodology to be followed by the development teams. The methodology included QA's involvement throughout the software development phases; however, the project manager chose to deviate from the methodology and excluded QA from requirements changes and design discussions. As the senior quality assurance analyst, I was provided access to the prototype that had been developed by a consultant that was no longer with the company. A Human Factors resource was my point of contact to address questions that I had regarding changes to the business and design requirements. I was responsible to author a test plan based on the portfolio's test plan template and author test cases that would validate the requirements. The prototype was the driver for building of the test cases, and any updates were provided by the Human Factors resource. (Human Factors professionals are trained in some combination of experimental or cognitive psychology, physiology, and engineering—typically industrial, mechanical, electrical, or software engineering. Human Factors engineering seeks to ensure that humans' tools and environment are best matched to their physical size, strength, and speed and to the capabilities of the senses, memory, cognitive skill, and psychomotor preferences. These objectives are in contrast to forcing humans to conform or adapt to the physical environment.)

A test concern was the data that was being converted from a mainframe database to an Oracle database. The conversion was handled with the development of wrappers that converted the data, and then the converted data was populated in the Oracle database. The validation of the data was not a concern of the lead developer or the project manager. I moved forward with planning the data validation and worked with one of the end users to determine what data should be populated in the database when specific functions were executed.

At test time, a significant number of functionality defects were being reported, and the data that was displaying in the application was in error. The data that was displayed was truncated, there was duplication of rows of data, and there was missing data. The discovery of these data issues were a result of the wrappers that were written to convert the data were not designed correctly, and in some instances the wrong interfaces were defined by the technical team.

In essence, the development team did not follow the methodology that would have involved the inspection of the code to ensure the software code that was written was defect ridden. The lack of a consistent structure wound up costing the business a significant amount of lost time and money to create a product that did not work, and the project wound up being shelved.

Obviously, there were a number of issues with this project, and if the development team would have followed the portfolio's software development methodology and if the project manager would have placed a higher value on the importance of quality best practices would have had a better chance for success. Certainly, a communication process would have been more fruitful to the project. If appropriate verification and validation activities would have been implemented before the project went into system test by the QA team, a number of the major problems and defects could have been avoided and the cost of quality reduced.

On behalf of the portfolio, the latter project was not the norm for following the methodology, as other projects that I was a team member of did support and follow the governance and standards established by the business. This example was to convey that when a business has a methodology, governance, and standards, the project manager may choose not to follow the process, and the result is that unfavorable events occur.

Testing does not just include the exercise of code or the execution of functionality to determine if the application is working according to the requirements. Testing also covers inspections and review of artifacts to resolve defects that may reside in the artifact.

The IT Metrics Productivity Journal and Special Edition, July 2005, documents an interview with Capers Jones, Chief Scientist Emeritus, SPR. Mr. Jones states:

> Sometimes companies go backward as well as forward. It's been kind of a drunkard's walk. What happens is that the generation that developed these things—the Zachmans and Watts Humphreys and so forth—they reach retirement age and go away. And the younger generation does not always pick up those solid lessons. So a lot of things being used by companies fifteen years ago with some success have dropped out of use while new and more ephemeral things like agile methods have begun to surface. And these newer methods sometimes work and sometimes don't. But by the time the latter is discovered, the people that knew that did the work have retired and gone away. Companies then have to relearn.[12]

In this chapter we have addressed the various Lifecycle models; however, there is no specific model that provides a view of high level test processes. *Figure 2.6 The Test Readiness Model* provides a high level perspective of the processes for test planning and test execution.

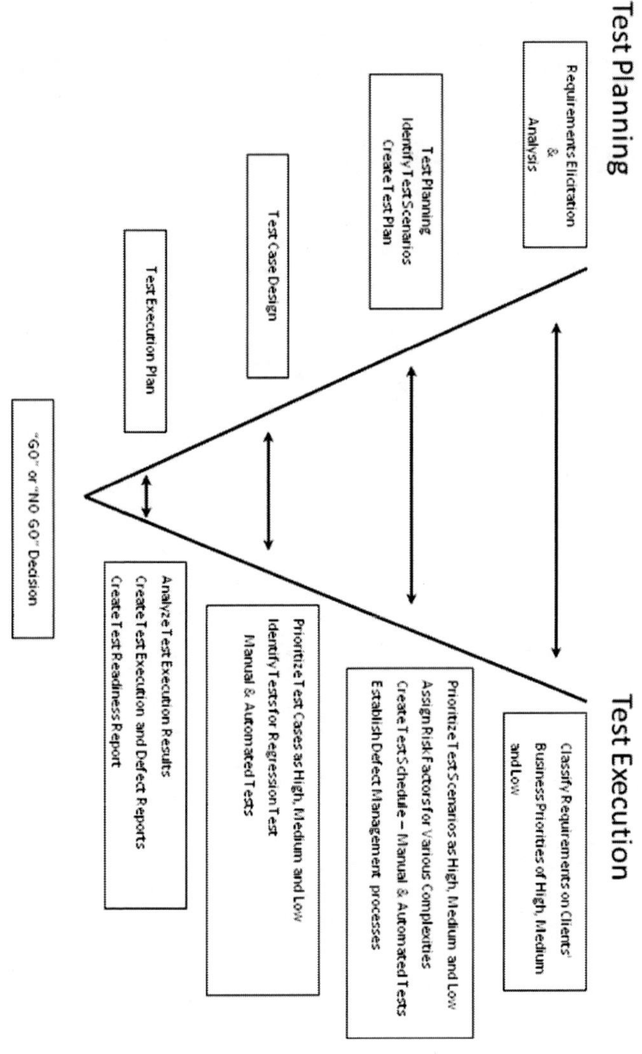

Figure 2.6 Test Readiness Model

The test planning and test execution sides of the model describe the processes that need to be in place.

COMMANDMENT THREE:

Do Not Eliminate Processes and Tests to Save Time or Resources; Testing Shall Occur in Each Phase of the Project

Testing is a very important part of the SDLC. However, I have found in my career that some companies do some unusual planning that does not follow best practices. Before I go forward with this discussion, the standard types of tests that occur for software development are:

- Unit Test is the testing of individual hardware or software units or groups of related units by the developer of the software code.[13] Unit Test is performed by the developer to ensure the logic they have programmed is functioning according to the system design specifications. Unit testing includes code coverage, branches, paths, statements, and the internal logic, etc. Unit Test is also known as "white box" testing.

- Prototype Test–This type of testing is testing a working model of the product that is built based on the design specifications. The prototype is generally limited in scope in functionality, and in general it is a limited version of the software product. In my past experience, not all prototypes have been working models, but some of them have been a composite of screen prints or wire frames that are reviewed or tested. The prototype is an opportunity for the business stakeholder to have an idea of what the product will look like.
- Gray Box Testing is a combination of white box testing (unit testing) and black box testing (system testing). The tester has to have more knowledge about the code and be able to perform some backend component testing. It is not uncommon for gray box testing to include testing of the prototype.
- Integration Test is the testing when the software components, hardware components, or both are combined and tested to evaluate the interaction between them.[14] Integration testing is usually performed when unit test has completed and the programs and modules are combined or concatenated. The develop-

ment team then executes tests to evaluate the interaction with the various components of the software product. Prototype test and Gray Box test may also occur during Integration Test.

- System Test is usually conducted on a complete, integrated system to evaluate the system's compliance with its specified requirements. Performance, Regression, and the ILTY tests are generally performed during the System Test Phase of the project. [15]
- Regression Test - This type of testing is selective retesting of a system or component to verify that modifications have not caused unintended effects and that the system or component still complies with its specified requirements. [16]
- Performance Test–This type of testing is conducted to evaluate the compliance of a system or component with specified performance requirements. [17]

ILTY Tests–these types of tests involve:

- Auditability–the ability to inspect or review an artifact to compliance of governance procedures, specification, contractual agreements and other established criteria
- Availability–the degree to which a system or component is operational and accessible

when required for use. Often expressed as a probability.[18]

- Compatibility–(1) the ability of two or more systems or components to perform their required functions while sharing the same hardware or software environment. (2) The ability of two or more systems or components to exchange information.[19]
- On a project for an airline, it was necessary to perform this compatibility test with every new application that was being moved into the production environment to ensure that the new application would not affect other applications the dispatchers were using. As we had discovered on a previous application that was moved into the production environment, the application affected the ability to view the weather application, which was essential in their process of planning a flight plan. Needless to say, the new application had to be backed out and the defect had to be resolved very quickly.
- Maintainability–the ability of ease with which a software system or component can be modified to correct faults, improve performance or other attributes, or adapt to a changed environment.[10]
- Reliability–the ability of a system or component to perform its required functions under

stated conditions for a specified period of time.[21]

- Recoverability–the ability to restore a system, program, database, or other system resource to a state in which it can perform required functions.[22]
- Scalability–the ability to execute a performance test that focuses on ensuring the software application of a software application that is under test can handle the increases in the workload. (A workload may represent an exponential number of resources processing transactions in a period of time.)
- Usability–is the ability of ease with which a user can learn to operate, prepare inputs for, and interpret outputs of a system or component.[22] It is not uncommon when usability testing occurs that the user does not have a help facility to aid the tester as it tests the application. In some instances, this test is conducted in a lab where the tester is viewed by observers through a window but testers are unable to view the observers.
- Acceptance Test–This type of testing is formal testing conducted to determine whether or not a system satisfies its acceptance criteria and to enable the customer to determine whether or not to accept the system. (2) Formal testing is conducted to enable a

> user, customer, or other authorized entity to determine whether to accept a system or component.[23]

A number of years ago, at the Atlanta Quality Assurance Association (AQAA), the meeting's program were to address the various definitions for testing. It was discovered in that meeting there were many different naming conventions that were used for system tests. Therefore, I have chosen to use the definitions provided by IEEE for most of the test types.

When testing does not follow the standard process as described by the Waterfall method, gaps occur. Working as a consultant for a company that is highly regarded in the industry for defect prevention and SIX SIGMA, the acceptance test occurred before the system test (Note: this also occurs in practice when a business is using the Agile model.) This practice of having the customer test the software before system test has been executed is asking to have a customer be unhappy with the product that is being delivered to them, as the customer is most likely encountering a significant number of defects. Why present a product that will leave a poor impression with the customer, or in a worst-case scenario, lose the customer because the product does not meet their requirements? In my opinion, it just does not make sense. When I encountered this process, I recommended to management they needed to change the process and perform system testing before acceptance testing. That way the defects that are encountered are resolved and the customer has a better impression of the product.

Another worldwide company with offices throughout the United States and international operations chose to perform the acceptance test while system integration testing was occurring. In looking back at the issues and the defects that were reported for the project, a number of factors caused a delay in the project. The development team was hampered with not knowing which defects to fix first, as the user acceptance team and system integration teams were filing defects at the same time and in some cases the same defect. Most certainly, defects that were defined as critical had to be resolved right away. What if two defects with severity level of Critical are filed at the same time by each team? (Defects defined as Critical must be resolved immediately, as usually this type of defect prevents usage of the software product.) This is where the problem occurred, as to which one to fix first. The vice president over the user acceptance team was of the opinion their defects needed to be fixed first. Additionally, he refused to have user acceptance testing after system integration test, which became a political power play. The user base was encountering defects that system integration had not fixed and was viewing the product before it was ready for the user to use as a business solution. Therefore, defect resolution resulted in a time consuming activity.

- The user had not been trained in defect management

- The user was able to view an application that was not completely ready
- The user was not versed in reporting of defects
- The user was reporting defects that were not repeatable
- The user was reporting duplicate defects
- Management team for acceptance test expected their defects be resolved prior to defects reported by the system test team.
- Managing the deployment of the fixed defects into a manageable schedule, and
- The system test schedule was impacted due to the time being allocated to resolve defects for acceptance test.

The schedule was not reduced with both tests occurring at the same time.

Unfortunately, the decision that was made to perform acceptance test at the same time as system integration test was made by executive management. Discussions were held to address this issue; however, the decision to go forward with both tests was made at the executive level.

In the planning for testing, resources need to be planned, and this includes testers, permanent staff, and contractor staff, data and hardware needs. If it is a new application, the data may have to be created. However, if it is a replacement application for a legacy type system, will the legacy type data be useable or will the data need to be converted?

Newly acquired hardware needs to be tested prior to a system test's start date to ensure there are no hardware problems. Generally, performance testing occurs in the system test phase; however, it may be wise in planning for the performance test to provide some benchmarks. A benchmark is a point of reference where something may be measured. For example, have some performance tests run against the legacy hardware and then when the performance tests are executed on the new hardware, the results can be compared. It may be prudent to run some performance tests in the integration test phase against the new hardware to ensure its capability and availability for when system testing is performed.

If there are other hardware needs, such as workstations, laptops and/or printers for system test, all of this hardware needs to be identified, and all the hardware will need to be configured for the test to be performed.

One area that is overlooked sometimes is the printer setup and to have a printer that would meet the business requirements. I was on a project for a claims management application, and there was a business requirement the claim's checks needed to be printed on a MICR printer. With that consideration, this required some security planning to ensure only specified resources would have access to the printer location and to validate the checks once the tests were executed. Therefore, when planning for system testing, the QA analyst would need to determine if there were some tests that would need to be performed other than testing of the application and the database.

If testing is for the Web, all the operating systems and browsers will need to be available to perform the test. In order to test this criterion, this may require a number of workstations be available for all the operating systems and browsers to be installed on these computers. Some operating systems are not compatible with all the browsers, such as Mac with Internet Explorer; therefore, if this is needed, allow for the additional workstations to manage this test effort. A heads-up here: it might be wise to label each workstation with the operating system and browser that is installed to ensure the tester knows exactly the set up for test they are executing.

However, there is an alternative for testing of multiple operating systems and browsers on a numerous workstations. There are imaging tools that create an image of the browsers and operating systems. Depending on the operating systems and browsers that will be needed for the test, this may require more than one work station; however, the number of pieces of hardware can be significantly reduced.

After the operating systems and browsers are imaged on a workstation, the testers execute a test script against the operating systems and browsers, such as Vista and (Internet Explorer) IE 7.0, and then execute a test script for Vista and IE 8.0. Testers and administrators need to know that these tools do not automatically execute for all the operating systems and browsers at the same time;

they must be executed separately. The savings here is that numerous workstations will be reduced.

There are a number of new imaging tools on the Web, and I suggest researching them for the tool that would fit your budget and testing needs. When I was searching the Web in the summer of 2009, I found some great tools for testing the Web and some new ones are on the horizon. In my search, I also found some that were open source freeware; however, there were some limitations in their capabilities for what may be needed for Web testing. When considering purchasing an imaging tool, there needs to be some additional time for training.

Offshore resources have become a player in system testing, and many companies have employed offshore teams. In many instances, they are successful, and in some instances they are not successful due to some of the reasons documented in Commandment One. A concern I have is that some CMMI Level V offshore teams do not always perform at that assessment level. This I have experienced on more than one project and it is very troubling when this occurs. I have informed the team that my expectations are for them to perform at their assessment level.

The other side of the coin is the test organization may not have been part of the CMMI Level V assessment, and when that is true, the test organization needs to inform the company that has secured them that they were not part of the CMMI assessment and may not perform at that level.

Of course, there are many questions that need to be posed for bringing on an offshore test team; however, there a few thoughts that needed to be considered in partnering with an offshore team:

- Find out if the test organization was part of the Level V assessment
- If they are not, then this needs to be conveyed to the management team that they will be their partner, and
- If they are not a Level V, was the test organization assessed at another level or not assessed at all.

From the business perspective:

- Is the company saving money with an offshore team that is producing work that results in constant rework?
- How much as a business are you willing to absorb in the cost of quality when there is considerable rework with an offshore test team?

I have managed QA offshore teams in India, China, and Mexico, and the most successful test was when I was involved in an implementation of a claims management application that included the testing of two hundred interfaces. This was a huge project that was approximately a forty-million-dollar project. The offshore team was from India, and the model that was effective in utilizing offshore, onshore, and company QA resources is documented in the *QA Offshore Model.*

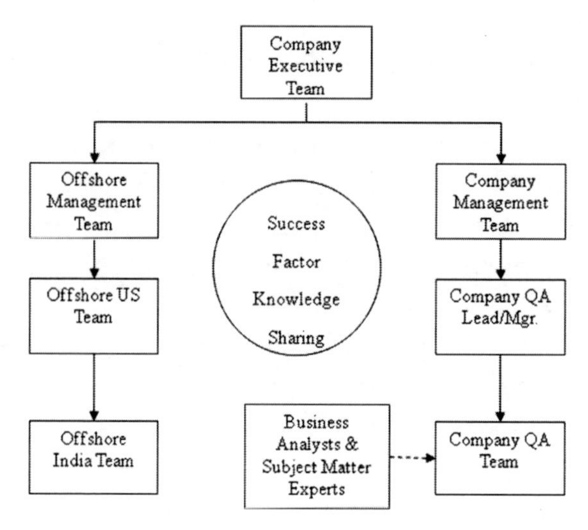

Figure 3.1 Offshore and On Shore QA Model

The success factor with an offshore QA model was the inclusion of a company QA team that provided knowledge sharing throughout the test phase of the project. The company QA team was supported by business analysts and subject matter experts. The offshore team was responsible for developing the test plan, the test schedule (this was a shared activity with the company QA lead) and the test cases. Initially, the company QA team was to provide knowledge sharing sessions with the offshore team, and the offshore team was to move forward with test casc development and test execution without the benefit of the company QA team. However, that plan was adjusted to include the assistance of the company QA team, as the knowledge and experience of the company QA team was needed to review the products created by the offshore team.

The company QA team was to review the test conditions and then follow through on reviewing the test cases. This resulted in a discovery that the offshore team did not have all the understandings or business knowledge of the business discipline and needed the company QA team in test condition definition and refinement to ensure the test conditions would validate the requirements. In reviewing the offshore team's test cases, it was also discovered that test steps were flawed, and this was due to knowledge gaps in the business discipline and application level knowledge.

My opinion is that when a company is considering the use of an offshore team for the test phase of a project, it should to proceed with caution and investigate the company completely to determine their success ratios in software testing. Don't let the reduced resource costs sway your decision, as it may cost the business more with additional rework.

Another area that needs to be clearly documented is the number of resources that will be used for the entire period of the project, even if the project ends up being extended beyond the planned end date. The business that plans to bring on board an offshore team shall request in their proposal the names of the resources that will be on the project, and any changes in number of resources and names must be approved by the company prior to a reduction in staff by the offshore company.

A USA company was using an offshore team for the test effort, and the project was being extended due to the number of defects that were being reported and the amount of time it was taking for defect resolution. In the daily management meeting with the offshore management team, it was discovered that the offshore test team had significantly reduced the number of test cases that were being executed. In addressing this concern, it was apparent the offshore team had reduced the number of resources significantly by their executive management. This decision by the offshore team's executive management had not been conveyed to the company's executive management.

In a tight economy, and when a project has a limited time to develop a product for the customer, it is not uncommon for management to request what can be done to reduce time for a project to get it out the door. When I have heard this, I've always cringed because it was not uncommon for some artifacts not to be completed to the detail that was needed for the project, and/or the time allocated for testing is reduced.

In this situation, the QA analyst needs to consider what can be done to test the product and reduce the probability of defects being delivered to the customer. One method that is effective is risk-based testing. There are those who believe that risk-based testing is a dangerous avenue to tread, as there may be tests that are not executed that will have a negative impact on the quality measurement of the application. To implement a risk-based test, a complete understanding of the process is necessary and must be agreed upon by management.

According to the Certified Quality Analyst (CQA) Examination Study Guide, risk is the probability an undesirable event will occur.[24] One of the many techniques in performing risk analysis is the "three key elements that must be considered":

- The cost of impact for when the event occurs
- The estimated frequency of occurrence or the likelihood of the event
- The risk expectation

The risk expectation (RE) is generally determined by a group of individuals from various departments, such as business, operations, development, network, and others. The risk process is:

1. Submit the risk
2. Assess the risk
3. Evaluate the risk
4. Mitigate the risk
5. Complete the risk

Unless the QA analyst is cognizant of what the risks are to the software application and to the business, the initiation of risk-based testing may be fruitless. The business and design requirements need to include what the impact would be to the business if a requirement is not delivered. The impact may be defined as high, medium, or low, and with this assessment the QA analyst will then be able to plan a risk-based test to reduce the probability of delivering a software product that will have a negative impact to the project and to customer's business need. Then, once those tests have been defined, the QA analyst will be able to determine if the timeframe will have enough spread to execute other tests.

In interviewing candidates for a senior level QA analyst position, I have found when I pose the question, "What is risk-based testing?" very few have been able to define what risk-based testing would represent or to plan tests for risk-based testing. Certainly companies that are hiring or promoting test analysts to the senior level should understand that the QA analyst or tester must be able to understand the importance of risk-based testing, and the analyst at this level must be able to build a test based on risk-based testing.

COMMANDMENT FOUR:

All Artifacts Shall Be Under Configuration Management's Control throughout All the Phases of the Software Development Life Cycle

This commandment is clearly stating that all artifacts shall be under configuration control, and this also means test cases or test scripts that are being created. In my twenty- five years of working in the discipline of quality assurance, there have been requirements that were drafted and not stored in the configuration management repository or in an automated test repository.

I was working on a project for a company that had a configuration management process; however, some of the resources chose not to follow those procedures. Requirements documents were stored on their work stations and were not downloaded to the configuration management tool. In other situations, requirements documents were sent out to the development team, only to discover the documents were not the most current version. When the QA analyst was validating the test conditions and test cases for the project, she discovered that the conditions and tests were not current, as it was determined there were more current requirements that resided on a business analyst's hard drive. Obviously, this caused additional delays and rework for the test and development team.

In this circumstance, the test cases or test scripts were not stored in the configuration management tool. I have found in reviewing test repositories where the test cases and scripts are to be deposited in the facility, there needs to be more due diligence in the management of repositories and in setting up the file structure for the viewer to easily determine that the tests support the requirements and the type of tests that were approved for the project and the respective release.

So often the testers who are writing the test scripts and storing them in the test facility fail to remove bogus test cases and/or trial test cases, which can confuse the reviewer of the test repository. In the planning process, there needs to be a resource that has the responsibility to ensure the test cases and scripts that are stored are valid and do not contain test cases/scripts that are trial or bogus tests.

The fundamental purpose of configuration management (CM) is to establish and maintain the integrity and control of software products throughout a project's life cycle. These products represent artifacts that are created by the business, development, database administration, operations, network services, and quality assurance to ensure the most current version of the artifact is being used and stored in a safe environment for retrieval during the development process.

The software CM process is comprised of the following integrated activities:

- Configuration identification of artifacts/work products used or developed by a project
- Configuration change control of information, including the impact of changes to organizations, management practices, schedules, budgets, technical or assurance activities, testing or retest requirements or project status
- Status accounting of artifacts/work products used in the development, release, and maintenance of a project
- Configuration reviews and audits that assess the status and acceptability of products controlled or released by CM
- Project delivery and release management procedures, and the capability to monitor the status of project information
- Establishment of a Software Development Library (SDL) and maintaining the integrity of the work products placed under CM control to ensure repeatability of the products and baselines.[26]

Inadequate configuration control is caused by inadequate capital investment, inadequate programs of study for management, curricula, inadequate software engineering educational programs, and inadequate standards.

I have discovered in years in the QA discipline that many of my colleagues are not cognizant of what a configuration item may be, and according to the IEEE, a configuration item (CI) is an aggregation of hardware, software, or both, that is designated for configuration management and treated as a single entity in the configuration management process. Configuration items may consist of:

- Plans
    - Systems Engineering Management Plan
    - Software Development Plan
    - Software Standards and Procedures Manual
    - Software Configuration Management Plan
- System Specification
    - Software Requirements Specification
    - Graphical analysis modes
    - Process specification
    - Interface specification
    - Prototype(s)
    - Mathematical specification

- Design Specification
    - Data design description
    - Architectural design description
    - Module design description
    - Interface design description
    - Software design folder
    - Objects description (object-oriented)
- Source Code
    - Build scripts
    - Source code listings
    - Executable programs
    - Module executable code
    - Linked modules
    - Legacy code (i.e., COTS, Government-Furnished
    - Information (GFI), Non-Development Item (NDI),
    - and so forth
- Operation and Installation Manuals
- Test Specification
    - Test plan and procedure
    - Test cases and recorded results
- Database Description
    - Schema and file structure

    - Initial content
- Software Description Document/Version Description Document
    - Compile and build instructions
    - Compiler(s) or network environment
- COTS products
    - Compilers
    - Computer-Aided Software Engineering (CASE)
    - Products
    - Models
    - Simulation software[27]

All of the above CIs are very important for the development of a software product. However, I have found in many places of business the only artifact that is stored is the source code. In performing my tasks as a QA consultant, I have found there are differing opinions in the business environment about what constitutes configuration management control. What surprises me the most is when a storage facility that is not a standard configuration management tool is being used as the repository for company assets? I was working on a project for a startup business, and one weekend, thieves broke into the business and pretty much wiped everything out of the business. Most of the artifacts were gone that weekend. I realize this is a worst-case scenario; however, if proper controls had been in place, most likely the artifacts would have been available.

On a project for a financial institution, the configuration management tool was for code. They were using a storage facility that did not provide a date for when the artifact had been stored in the document repository or what was the latest version. The only method of determining the latest version was to open the document and review the document history. The most surprising was the project management office supported this method. At one point in the project, someone deleted all the files, and I do not need to say anything more of the uproar this caused. Still, they would not consider using a configuration management tool for all the business assets.

Another situation a number of years ago: I was on a project that was shelved; however, years later I was contacted to find out where the test plan and test cases were located, as the project was activated again and they needed to know where the test plan was located to build the test environment. Unfortunately, the version that I had checked into the configuration management tool was not the most current version, as another QA analyst had taken over the responsibility from me and had not checked in the most current version of the test plan. The QA analyst had left the company, and I had transferred to work on the Y2K effort for the company.

I informed the company I understood a hard copy was stored in a black binder that had been placed in the test lab; however, that copy was also missing. What is important here is that company assets need to be checked into the configuration management tool and remain checked in. The configuration management tool allows a resource to check out a copy of the artifact, and thus the individual can update the document. With this process, a version still remains in the repository and avoids the problem of losing a business asset for the company.

To avoid improper handling of source code and business assets, it would be wise for a business to bring on board an individual who is responsible for managing the configuration items and providing the reporting that is needed for a project. Configuration Management Analysts (CMA) are responsible to develop, implement, maintain, and document all tools to support the configuration management process. The CMA creates and maintains scripts, procedures, and documentation to support the building process and the product build/release and software administration for the project's environment.

Additionally, the role may require the CMA to provide:

- Status accounting of artifacts/work products used in the development, release, and maintenance of a project
- Configuration reviews and audits that assess the status and acceptability of products controlled or released by CM

The role of the CMA may include other duties as required by the business to support the configuration management process.

COMMANDMENT FIVE:

Testing Shall Always Be Planned

I was heading home through the Smokey Mountains on a beautiful autumn day in 1992 when I noticed a crowd had formed at one of the overlook locations. I had to find out what all these people were looking at, and much to my amazement, there was a brown bear sleeping in the crotch of a very large and tall tree. At the base of the tree, a photographer had set up the tripod and camera to capture the brown bear resting. If that bear had decided to quickly leave his place of resting, the photographer would not have had time to escape and get to a place of safety.

Obviously, the photographer had not planned for an unfavorable event to occur. Well, planning is also very important for testing of an application to avoid unfavorable events when all the necessary tasks are not in place.

Before test execution is to begin, there are test artifacts that need to be created, reviewed, and approved before the test phase start date. These artifacts are: a detailed test project plan for the test phase, a test plan, a traceability matrix, a test schedule, test cases, and a communication plan. The communication plan, test schedule, and traceability matrix may be included in the test plan, as an appendix in the test plan or as separate company assets. The detailed project plan for the test phase shall include the planning for all the milestones, test tasks, deliverables, resources, dependencies, and test constraints.

When I began my career in QA, all I knew was the business side, and I utilized that business knowledge when validating the developer's logic. At that time in my career path, I had no concepts of process improvement, IEEE standards, test plans, checklists, or any of the test artifacts that are successful and helpful in planning of a test phase, let alone any considerations for planning of a test phase. My artifact was a hard copy of an insurance manual's state rating pages and exception pages, which was used as the requirements document to write manual test cases that would validate the application was working according to the requirements.

Over the years of serving as a software tester (test engineer), QA lead, and QA manager, I created my own toolbox that has assisted me in managing and planning for test execution. The project needs determine which tool I will use and when. One of my favorite tools is a checklist, and it is considered one of the most valuable tools for quality assurance. I have used checklists in the planning efforts for test execution. To begin the process of creating a checklist, Excel has been my software tool of choice, as I can sort and filter as I progress in writing the checklist.

For the readers and testers who are looking for an example of a checklist, I have provided an example of a checklist that is structured for test planning and test execution. This checklist is not all inclusive of all the criteria for test planning and test execution, as this depends on the project needs; however, it does provide a good starting point for the QA resource who is looking for a tool to assist in his or her QA planning efforts for test planning, test case development, and test execution. Please note the checklist *is not targeted toward a specific technology*, and criteria would need to be added for specific technologies or test types.

Checklist for Test Planning and Test Execution					
Number	**Item**	**Yes**	**No**	**N/A**	**Comment**
	General Planning Checklist Items				
1	Have resources been assigned to the business when requirements are being gathered?				
1a	Have requirements been reviewed?				
1b	Have the requirements been approved?				
2	Have all the design artifacts been reviewed?				
2a	Have all the design artifacts been approved?				
3	Is there a defect log that documents the defects found during the requirements review?				
3a	Is there a defects log that documents the defects found during the design phase?				
4	Is there a planned Handover Review in the Project Plan for entry into the test phase of the project?				
4a	Did development provide a list of outstanding defects on other test information that is vital for the QA Team at the Handover Review?				
5	Has a To Do list been created for the test?				
6	What method of tracing the requirements to the test cases/test scripts will be used?				
6a	Will there be an automated tool used for traceability?				
6b	Manual, such as Excel spreadsheet?				
6c	Have all business requirements been traced?				
6d	Have all design artifacts been traced?				
6e	Have all changes to the requirements been traced?				
6f	Have all use cases been traced?				

Checklist for Test Planning and Test Execution					
Number	**Item**	**Yes**	**No**	**N/A**	**Comment**
6g	Have all test cases been traced to requirements?				
6h	Have all test cases been traced to use cases?				
	Test Planning				
7	Are there templates to use for test planning for:				
7a	Test Plan?				
7b	If there is a master test plan, does the project test plan align with the master test plan?				
7b	Test Cases?				
7c	Test Schedule?				
7d	Communication Plan?				
8	What type of test will be executed, such as system, performance, regression, acceptance, etc.?				
9	What is the objective of the test?				
10	Is there a corporate test strategy?				
10a	Have you defined the test strategy?				
11	What will the test strategy address:				
11a	Unit Test / White Box Test?				
11b	Integration Test?				
11c	Gray Box Test?				
11d	Smoke/Sanity/Dry Run Test?				
11e	System Test?				
11f	Regression Test?				
11g	Performance Test?				
11h	Scalability Test?				
11i	Usability Test?				
11j	Reliability Test?				
11k	Compatibility Test?				
11l	Availability Test?				
12	What is the scope of the test?				
12a	What is out of scope for the test?				
13	Has the entry criteria been identified that will be needed for test planning and test execution?				

Checklist for Test Planning and Test Execution					
Number	**Item**	**Yes**	**No**	**N/A**	**Comment**
14	Has the exit criteria been identified?				
15	Have you identified test assumptions?				
15a	Have you quantified the assumptions as to the per cent of probability that the assumption is incorrect?				
16	Have the critical success factors been identified that would provide an indicator the test is completed?				
17	Has the suspension criteria been identified to determine what factors would result in stopping the test?				
18	Have risks been identified that would affect the overall success of the project?				
18a	Has an impact analysis been performed for the documented risks?				
18b	Has a mitigation plan been documented for each risk?				
18c	Has a contingency plan been documented for each risk?				
19	Have completion criteria been addressed?				
20	Have resource needs been identified for the project, such as:				
20a	Number of resources needed to execute the test?				
20b	What type of skill sets be needed for the test been considered?				
20c	Will you need full time employees?				
20d	Will you need contractors?				
20e	Are there hardware needs?				
20f	What data will be needed for test execution?				
20g	Will there be a need for support services from other project teams?				
20h	Will there be training needs?				
21	What is needed for the test environment?				
22	Will automated tools be used for testing?				
22a	If yes, what tools will be used?				
22b	What type of automated tool will be used for the test type?				
23	Will there be a need to access the database?				

Checklist for Test Planning and Test Execution					
Number	**Item**	**Yes**	**No**	**N/A**	**Comment**
23a	If yes, has access rights been established for the database?				
24	Have testing priorities been addressed for the test?				
25	Will there be a test run (sanity or smoke) test be executed?				
25a	If yes, who will have that responsibility?				
25b	What is the timeframe for the test run to be executed in (hours) (days)?				
25c	What test cases will be executed for the test run?				
26	What test artifacts will be delivered in the test phase?				
27	Has a RACI (Responsible, Accountable, Consulted or Informed) been created to identify resource roles and responsibility for the test phase?				
27a	Have all the roles been identified that is to be listed on the RACI chart?				
	Test Case Planning				
28	Do the test cases support the standard test case template?				
29	Are requirements/use cases referenced in the test cases?				
30	Are test priority documented in the test cases to determine when the test case(s) are to be executed?				
31	Are risks documented in the test cases to determine the impact if the test case fails?				
32	Are predecessor and subsequent test cases documented on relevant test cases?				
33	Are dependencies documented in the test cases that identify a dependency must be available to execute the test case successfully?				
34	Have the test cases been reviewed?				
34a	Have the test cases been approved?				
35	Are regression test cases identified?				
35a	Is there a plan as to which test cases will be executed for regression test?				
36	Have test cases been created for				

Checklist for Test Planning and Test Execution					
Number	**Item**	**Yes**	**No**	**N/A**	**Comment**
	requirement changes?				
36	Are data needs documented in the test case?				
36a	Have discussions occurred with database administrators for data needs for test execution?				
	Test Schedule Planning				
37	Has a test schedule that supports the test schedule template been documented?				
37	Does the test schedule allow for :				
37a	Test case number?				
37b	Test Case Description?				
37c	Cycles?				
37d	Priority?				
37e	Dependencies?				
37f	Planned Execution Date?				
	Test Execution				
38	Is software base lined prior to test execution?				
39	Is there a communication process to inform the test organization and other teams the environment is ready for test (this should be included in the Communication Plan)?				
40	Have all test artifacts been approved prior to test execution?				
41	Have it has been determined who will lead the Daily Test Meetings?				
42	Has it been documented, which tests that support the business requirements will be executed manually?				
43	Has it been documented, which tests that support the business requirements will be executed with an automated test tool?				
44	Are there batch jobs that will be executed during system test?				
45	Will data need to be transmitted in an outbound interface?				
45a	Have the resources been identified to validate the outbound data?				

Checklist for Test Planning and Test Execution					
Number	**Item**	**Yes**	**No**	**N/A**	**Comment**
46	Are there inbound interfaces that will need the data valid?				
46a	Have the resources been identified that will be validating the inbound data?				
47	Will a calendar need to be created to determine when an interface will be tested?				
48	Has a meeting been planned to discuss the interface test process plans for outbound and inbound interfaces?				
49	Have test executions reports been identified/created that will report the test case and test script test execution results been reviewed and approved?				
50	Is there a defect management tool to report defects?				
50a	Have all resources been setup in the defect management tool?				
51b	Have defect severity levels been agreed upon?				
51c	Will any resources need to be trained on the use of the defect management tool?				
52	Have defect reports that will report the status of defects outstanding been agreed upon and approved?				
52a	Have defects reports been identified to disseminate to management?				
53	Is there a plan for defect meetings to address defect resolution and assignment?				
54	Have reports been prepared for the Go/No Go Decision?				

Figure 5.1 Checklist

A test plan that is created in an automated tools repository is not a test plan that would document all the planning that is needed for a test. The plan that is built in the automated tool is a tree structure of the tests that supports the requirement's functionality. In interviewing candidates, I have found that most test engineers who have written test scripts in an automated tool are of the opinion this represents a test plan. Unfortunately, that is not always the case and the tester needs to understand the difference between a test plan that is created in an automated tool's repository and a test plan that is a written as a word document that documents all the planning for test execution. If it is the opinion of the business that the test plan that is created in the automated test tool is adequate for test planning, then there is the probability that some planning needs may not be considered for test execution, which can lead to issues occurring prior to and during test execution, which may extend the test phase timeline.

The resource that is assigned the responsibility of authoring a test plan must have the characteristics of looking at the entire effort for test planning, and this requires an individual who is able to define all the needs for the test effort and who is also able to be more concrete in defining the specifics of the planning effort. As an example, the test environment section of the test plan would not just state the test will occur in the test environment, but the section would describe the hardware platform, database (such as Oracle 10i), and software elements, etc.

I served as a consultant for a staffing company, and in that position, I also interviewed QA candidates. My discovery was that when I would ask the question what a test plan should contain, unfortunately very few candidates could provide a laundry list of the primary sections of a test plan. This, in most part, was probably due to the fact that they learned their QA skills from the company they were employed by. If the company did not offer training programs in the quality assurance discipline, the QA resource was not fully knowledgeable in the skills to plan and perform testing and quality assurance activities.

Certainly, there should be some initiative in advancing his or her skills; however, the QA analyst will need someone to mentor him or her. Over the years, I have encountered time and time again QA directors/managers who have no previous experience in the discipline of quality assurance. This is the person who is going to lead the quality efforts. Come on, companies. Wake up and smell the roses. If you are placing someone with the responsibility of leading your quality efforts, at least hire the person who has strong QA skills and experience to lead the quality effort. Who is going to mentor the quality team if the leader has no experience in quality? I am not talking about someone who goes out on the Web and reads about QA; this person needs years of practical experience to lead the team.

I am always surprised by:

- How many QA professionals do not attend a QA professional organization to further their education in the quality discipline

- How many QA professionals think they know everything about QA and they do not
- How many QA professionals do not have a library of books that aid the professional about other approaches for testing and process improvement
- How many QA professionals have learned their craft in a company that does not offer training programs or that will not pay for classes where the QA analyst has an opportunity to learn the craft that will benefit his or her testing skills
- How many companies offer all kinds of training programs to developers but not to the QA resources

If a business does not support training programs for the QA professional, there are a number of professional quality organizations and quality conferences that offer programs whose purpose is to educate the QA professional in the discipline of quality assurance. These programs cross the gamut of process improvement to testing. Today, there are also webinars that are being offered, and a number of them are free to the attendees. QA professionals, it is time to for you to take the initiative and become educated in your role, and find out what is out there to advance your skills in the method of test planning and building a test plan that describes all the entities that are necessary for a successful test execution.

According to the IEEE, a test plan is a document describing the scope, approach, resources, and schedule of intended test activities. It identifies test items, the features to be tested, the testing tasks, who will do each task, and any risks requiring contingency planning.[28]

If your test organization does not have a QA test plan template, I have provided below an example of a generic QA test plan, and this plan is not specific to a test type. The following test plan is certainly not all conclusive of every section that shall be included for your organization; however, it is a beginning that may act as a template for your use. The plan does contain a section for test strategy, and if your company has a corporate test strategy, then you will need to ensure the test strategy section aligns with the corporate test strategy.

Test Plan

1. Introduction

Enter a written description about the product and what its intent is for the customer.

1.1 Purpose

Describe the purpose of the Test Plan for the specified test.

2. Test Strategy

Describe the coverage of testing that will be performed. As an example, test coverage will consist of:

- Functionality
- Regression

- Performance
- Usability
- Compatibility, and/or
- Web Specific Testing, i.e., html syntax checking broken links, interfaces browsers, graphics loading, etc).

2.1 Scope

Describe the scope of the project that will meet the customer's business and technical requirements that are critical to the life of the business.

2.1.1 In-Scope

Describe what is in scope for the <test type>.

2.1.2 Out of Scope

Describe what is out of scope for the <test type>.

3. Definitions

In this section, list all the acronyms that are in the test plan and provide a definition for the acronyms and other test terms that may need explanation to the reader. It is suggested that this information is set up in a table format for easy reading.

4. References

Document the artifacts that were used to prepare the test plan, such as business requirements, system requirement specifications and the location of where the artifacts may be found. It is suggested that this is set up in a table format for easy reading.

5. Entry and Exit Criteria

Describe the entry criteria and exit criteria for the <test type>.

5.1 Entry Criteria

Document the entry criteria for test execution to begin, as these documents support the test effort. The entry criteria shall include:

- Reviewed and approved business and design requirements
- Unit/Integration test results to determine test coverage
- Memory leak test results from the unit test execution memory/virtual memory loss
- Performance bench mark test results from the construction phase of the project
- Defect report from unit/integration test to determine test readiness

5.2 Exit Criteria

The exit criterion quantifies the status of <test type> testing and determines the readiness of the application to move to the next phase of the project. Examples of exit criteria are:

- 100 percent of the high priority test scripts that support the technical and functional risks were tested successfully
- 100 percent of the defects reported as "Critical" and "High" are resolved
- The customer has been consulted and reviewed the number of defects listed on the defect log and agrees that the application is production ready.

6. Assumptions

Describe assumptions associated with the testing project and quantify each assumption to the percent that the probability of the assumption is incorrect. An example of an incorrect assumption would be that the test environment is an exact mirror of the production environment. The probability of this assumption being incorrect is 70 percent.

7. Testing Dependencies and Constraints

Describe dependencies and constraints associated with the testing project.

7.1 Dependencies

Describe test dependencies associated with the test phase.

7.2 Constraints

Describe test constraints associated with the testing phase.

8. Suspension Criteria and Resumption Criteria

8.1 Suspension Criteria

Describe the criteria in which the test may be suspended. An example of a suspension condition is:

- Unable to access the application
- A critical defect has occurred and there is there is no workaround to continue testing, or
- Test cases are blocked from continued test execution, as unresolved defects are preventing continuation of the test.

8.2 Resumption Criteria

Describe the criteria that the testing is able to resume for the <test type>, such as a critical defect that prevented testing has been fixed and retested.

9. Risks

Describe risks associated with the testing, the mitigation and contingency plan to address the risk. In documenting the risks, the responsibility party for this task would:

- Describe the risk, such as the data for test environment is not representative of the data needed to execute all the test cases/test scripts
- Describe what the impact would be if not all the data is representative to execute the tests, and this is documented in a format of critical, high, medium, low or a numeric format

that the company has determined to use as their risk factors

- Describe the mitigation plan to prevent the risk from occurring, and
- Describe a contingency plan if the mitigation plan will not work.

10. Test Environment

This section describes the test environment that will be used to perform the actual testing activities. Any dissimilarity between the testing environment and the production environment are noted and reviewed to determine any impact. The entities that are to be described in this section are provided as examples:

- Hardware (servers, printers) needs
- Interface components
- Software system platforms and supporting applications
- Necessary tools and databases for logging, tracking, and reporting test results
- Supplies needed to perform the tests
- The testing facilities (e.g., Model Office, if applicable)
- Pre-deployment test, or
- User ID access.

10.1 Test Environment Name

Document the name of the test environment, the purpose for the environment, the physical location, the version of the environment, and additional information that may be needed. It is suggested this information is documented in a table.

10.2 Base System Hardware:

Document the base hardware elements that are required for the test, as this would include the name of the hardware element, a description of the hardware element, and the version of the hardware element.

10.3 Base Software Elements:

Document the base software elements that are required in the test environment. The information that will need to be documented is the name of the software element, a description of the software element, and the supported platforms. The example that is provided is for a Web site application:

- Software Element Name is "Application Server Oracle."
- Description is "BEA Application Server 10g v 10.3"
- Supported Platform–Operating Systems are: Windows 2000, Windows Server 2003.

10.4 Test Data

Document the test data that will be needed for the <test type>. Identify if the data represents a full set of data, i.e., production data, or if it is a representative sample of data. In planning your test data, it is important to ensure there will be adequate data to perform the test execution tasks. Specify any actions needed to handle personal and sensitive data if applicable.

10.5 Test Environment Build/Resources

Document the component that is installed in the test environment, the build number or version of the build, and who built the component. The reasoning behind this need is if there is a problem with executing the software, the build may have been defective; therefore, management is able to discuss with the resource the differences and resolve the build problem. After the build is completed, a test run will need to be executed to ensure that all aspects of the test environment are installed and commissioned ready for testing.

10.6 Test Tools

List all the test tools that will be used with a schedule indicating when they will be best used. These tools include the software, hardware, and system tools that will be needed to effect testing. Criteria for consideration are:

- Testing tools (e.g., test data generators, test drivers, etc.) installed
- Debugging tools
- Capture/playback tools and load testing tool, or
- Database, network, and performance monitoring products.

10.7 Training

Describe training, if any, required to install and operate the environment.

11. Test Resources, Training, Schedule, Deliverables, and Execution Approach

Describe the test resource roles and responsibilities, training needs for the testers, the test schedule, the test artifacts, and test execution approach.

11.1 Test Resources

Describe the test resources and their roles and responsibilities for the test phase of the project. A RACI Chart provides a high level example of the roles and responsibilities for each of the tasks that have an impact on the Test Phase of the project. See Figure 5.2 for an example of a RACI Chart.

RACI Chart **R – Responsible for doing the work** **A – Accountable for the delivery of the task** **C – Consulted during task execution** **I – Informed after the task is completed**									
Task	Project Mgr.	QA Mgr,	Bus. Analyst	Tech Lead	Operations	System Analyst	QA Lead	QA Analyst	Customer
Test Phase									
Create System Test Plan		A		C			R	C	
Review System Test Plan	A	R		R		I			
Create Test Cases/Scripts		A					C/R	R	
Review Test Cases/Scripts	A	A					R		
Execute Test Cases/Scripts		A					R	R	
Report Test Metrics to Management	A	A					C	R	I

Figure 5.2 RACI Chart

11.2 Training

Describe training needs for test resources.

11.3 Test Schedule

The test schedule identifies the tasks, resources, and milestones for the test initiative, and most often this information is documented in the project management tool. Reference the project plan and the location of the plan in this section of the test plan.

11.4 Test Deliverables

List all test artifacts and the location where the test artifact may be located.

11.5 Test Execution Approach

11.5.1 Test Priorities

Describe the test priorities for the specified <test type>. The test priority will determine the order the test cases/scripts will be scheduled during test. Priority is based on business need. An example of a test Priority for a Web test:

The test of Windows XP has priority over the test of Windows Vista.

11.5.2 Test Run

Describe the plans for executing a test run to ensure the environment is ready for test execution. This includes when the test will be executed, who will be executing the test, what will be tested, the expected amount of allocated time for a test run, and to whom the test results will be disseminated.

11.5.3 Test Execution Metrics

Describe the metrics that will be reported on a daily basis during test execution for the test, such as:

- Number of prepared test cases
- Number of test cases/scripts - Executed/Run
- Number of test cases/scripts - Passed
- Number of test cases/scripts - Failed, or
- Number of test cases/scripts not executed /run.

12. Defect Tracking and Reporting

12.1 Exception Handling

Describe the types of exceptions that may occur and the procedures to handle them. Include escalation and resolution procedures when applicable.

12.2 Defect Management

Describe the defect management procedures for the test.

12.2.1 Defect Status

The defect status and their respective description and definition that will be used during the test phase are:

12.2.2 Defect Severity

Document the severity terms and their respective definitions that will be used during the test phase:

12.2.3 Defect Priority

A defect priority is a guideline to determine the importance of correcting a defect based on the customer's need. Document the priority status and its respective definition.

12.2.4 Defect Metrics

Describe the defects metrics that will be reported daily during the test. Examples of these metrics are:

- Number of defects reported by severity
- Number of defects by status
- Average number of closed defects per day, by severity
- Average number of closed defects per day, by priority
- Defects by trend, and/or
- Defect aging report

13. Release Notes

Describe information that is specific to the release. This may include:

- Installation information
- Download information, if applicable
- Issues
- Outstanding defects

14. Test Plan Sign Off

Describe the sign off procedures (who, what, where, when, how).

Over the years of authoring test plans, I have always been surprised when the project manager requests the test plan to be written in its entirety before all the technology needs are defined. A good test plan is not written in a day or a week, as it will require discussions with all the teams that are involved in the project to properly define all the criteria for test execution to occur successfully. Generally, this document is an ongoing document from the beginning of the project to just prior to test execution.

Sometimes the resource responsible for authoring the test plan will have to stand up and hold his or her ground for an adequate amount of time to complete the test plan, and the project manager needs to be informed as to the importance of the test plan and all the criteria that needs to be defined for a successful test. As the test plan is being drafted, it is recommended to check the document into the configuration management tool as a "draft" document. This way, other departments can view it and identify areas that they need to be provided to the author of the document. Note: When the document is checked into the configuration management tool it is locked from other parties making changes to the document; however, it does not preclude them from making a copy and providing updates to you from the checked out copy.

Test Schedule

A Test schedule documents when the tests will be scheduled, who will be executing the tests, when the test is planned to be executed and documents the priority of the test to be executed and any dependencies that the test may require. In my experience, I have seen schedules that are so very finite that they describe the date, the day, and the time the test is scheduled. When I have reviewed this type of detailed schedule, I have to ask the question: what happens if there are defects on previous test cases where there is a dependency? I would like for you to think about this: if there is a dependency to another test case and the previous test case fails, then the dependent test case cannot be executed. The test is blocked from execution, and then the schedule is already in error.

I was on a project where the schedule was detailed exactly as described in the previous paragraph. Every test case was scheduled on a specific day, hour, and minute. Right from the onset of testing the software application, the test team encountered a number of problems. The test schedule was immediately off and had to be updated.

A word of caution in drafting a test schedule to the exact time every test case will be scheduled: defects do occur and this may affect the execution of a number of test cases. Unfortunately, the test analysts are constantly updating the schedule, which can become a very timely task. Certainly, the project would be better served to be able to allocate them to other tasks.

Generally, the test schedule is documented in the project plan and/or in a test management facility; however, in my career I have encountered the need for a test schedule that would document which test cases were to be scheduled, who would be executing the test case, what the priority of the test case was, and in which cycle the test case would be executed.

Test Case Number	Test Case Short Description	Resource	Priority	Cycle Number	Cycle Number ^{n}
1	Address field	M. Smith	Medium	1	
2	Transmit outbound interface data to ABC Insurance Company	T. Jones	High		2

Figure 5.3 Test Schedule Example

Additional criteria for the test schedule would be to add two columns, one for the date the cycle will be executed and a second column for dependencies. The dependency column would be populated with a resource that would need to be contacted to execute a function, such as a batch job. With the publication of a test schedule, any deviations from the proposed schedule will require the responsible resource to update the test schedule.

Test Cases

Manual test cases can be written in an Excel spreadsheet, Word document, or in an automated tool. Whatever form is used, a significant amount of planning, analysis, and organization needs to be in place to ensure test cases are created that support the approved requirements. When a test case is dependent upon another test case that is executed first, this must also be documented. I have found when I have reviewed test cases and there is a dependency to another test case, the dependency is not notated. The most common format of a test case is displayed below:

Window/Page/ Dialog Box	Test Case Number	Short Description	Step Number	Step Description	Data Input	Expected Result	Pass/ Fail	Actual Result

Figure 5.4 Example of a Test Case Template

Unfortunately the template that is provided as an example in *Figure 5.4* does not address traceability to a requirement, the priority of executing the test case, or the risk to the business if the test is a failure, and it does not provide a reference to a predecessor test case or a dependency to a subsequent test case. The format that would meet the objectives of requirement and dependency to other test cases is provided in *Figure 5.5.*

<table>
<tr><td colspan="2">Req't Number</td><td colspan="3">Requirement Description</td><td colspan="3">Priority
High, Medium or Low</td><td colspan="3">Risk
High, Medium or Low</td></tr>
<tr><td></td><td></td><td colspan="3"></td><td colspan="3"></td><td colspan="3"></td></tr>
<tr><td colspan="5">Dependent Test Case Number*</td><td colspan="6">Short Description</td></tr>
<tr><td colspan="3"></td><td colspan="2"></td><td colspan="6"></td></tr>
<tr><td colspan="3">Window/Page/ Dialog Box</td><td>Test Case Number</td><td>Short Description</td><td>Step Number</td><td>Step Description</td><td colspan="2">Expected Result</td><td>Pass/ Fail</td><td>Actual Result</td></tr>
<tr><td colspan="3"></td><td></td><td></td><td></td><td></td><td colspan="2"></td><td></td><td></td></tr>
</table>

Figure 5.5 Test Case Format with Requirement and Subsequent Test Case Reference

Dependent Test Case Number may require the entry of a predecessor test case with a subsequent test case reference.

Communication Plan

In the planning for the test phase, a lack of communication with team members and with other project teams may result in a less harmonious relationship with all the teams. The project team wants and needs to be informed when events are going to be communicated, what information will be communicated, and how they will receive the information. When information is disseminated to all management and non-management resources on the project, it stimulates an atmosphere of harmony.

Coming together is a beginning,
Keeping together is progress,
Working together is success.
Anonymous [29]

It is the responsibility of management to ensure there is a healthy communication environment for the project. How, what, and when the information is conveyed can aid in team members feeling the project is a partnership when information is disseminated to all the teams.

The communication plan is written as a formal plan with sections and resembles the structure of a test plan with an introduction, objective, etc. Another approach for a communication plan is in a matrix format, as in *Figure 5.6,* which is a very simple format for a communication plan, but does not provide an idea of the type of information that will need to be conveyed to the project team.

<Test Type> Communication Plan			
Document/Event/Task	**Method of Communication**	**Recipient**	**Frequency**
Daily Meeting	Email	QA Team Development Operations	Daily
Status Report	Report sent by email	QA Manager Project Manager	Weekly

Figure 5.6 Communication Plan

Whatever format is used for a communication plan, the document must be reviewed, updated, and approved.

To emphasize an important point, a communication plan is not a deliverable plan. I was on a project where a new manager was hired and was reviewing the communication plan. The manager determined that it was a deliverable plan. A communication plan is a method to inform all those involved in the project when various pieces of information are going to be conveyed and how they will be conveyed. The communication plan shall be reviewed and approved by the appropriate resources.

A deliverable plan documents the documents that will be authored and when those documents will be delivered. A deliverable plan is a business asset and must go through the process of being reviewed and approved. It should not be taken lightly and not be construed that it would fit the format of a communication plan. I was serving as a QA manager for a state government's child support enforcement program, and there was an in-depth deliverable plan that specifically identified the documents that would be delivered and when they were planned to be completed. This plan was reviewed regularly and validated on a regular basis to ensure the assets were being created for the planned due dates and the project staff were meeting these important deliverable milestones.

In the situation documented in previous paragraph, I made the mistake of not coaching the director about the difference between a communication plan and a deliverable plan. I did not have the mettle at that point to stand up and be counted. Therefore, when the responsibility falls to developing various business assets, the QA resource needs to coach management in the difference of a communication plan and a deliverable plan.

COMMANDMENT SIX:

Change Management Must Be a Process in the Software Development Life Cycle

Change management is a discipline of regulating the smooth flow of changes. It provides a system to properly govern changes in order for IT to achieve their business goals when change occurs. A change can involve any configuration item or an element of the IT infrastructure. Some types of changes are:

- Application changes
- Hardware changes
- Software changes
- Network changes
- Environmental changes
- Documentation changes

Change happens, and when it is not managed properly, it may cost business delays. An area that needs to be addressed is the planning for installing a change into the production environment and to ensure back-out procedures are in place in the circumstance when the installed change has a negative impact on the production environment.

I was managing a QA team for a dotcom company when we were informed by the development organization that they had converted the code from Perl to Mod Perl, and there were some business changes requested by the business stakeholders to the Web site. Meetings were immediately requested by the QA organization to first test the application for the modified code, then follow up with the tests for the changes to the business requirements. Our request was denied, and it resulted in:

- Many additional weeks of testing
- A significant increase in defects
- Eroded working relationships with the development organization, and the QA organization

Unfortunately, the business viewed the test efforts as costing additional time and effort that was unnecessary.

The owners of the business determined the efforts of the technology teams were not productive, and massive changes were implemented in the organization and eventually, the dotcom company was brought back under the umbrella of the organization that initially spun off the dotcom as a separate business. Obviously, this is a perfect example of when change management was not followed and the negative impact it can have on a business.

Ideally, a change management application would be utilized to track the changes; however, not all companies have a change management system. Then it is recommended the business create a change management process that will ensure the changes are following a logical flow for the implementation of a change.

It is not uncommon for the business to request a change in the requirements. When this occurs, the process to implement a change request is to:

- Plan a change
- Test and validate the change by assessing the feasibility of the change
- Create the change proposal
- Approve the change request, and if the change request is not approved it is returned to the initiator
- Document the request for change
- Create the change request
- Review the request for change (RFC). If the request for change is rejected at this phase of the process, it is then returned to the initiator
- Approve the request for change
- Update the project plan with the change request

- Update the configuration management database with the updated project plan
- Review and evaluate the RFC
- Update the change request log
- Update The Configuration Management database with the Status Report for the change request
- Update the configuration management database with the updated traceability matrix, and
- Implement the change request (configuration control system).

The QA organization is to be involved when the RFC is reviewed in order for them to have the understanding for test planning. To wait until the process is completed only adds more time for the test design to be implemented. After the change request is approved, the QA resources then draft the test cases that will support the change request. If the change is implemented after testing has begun, this will add additional time and effort by the test organization to draft the test cases, to plan when the additional test cases will be executed within the test schedule, and to determine the need for additional regression testing.

When the process of managing change is not followed, a number of scenarios may occur:

- The changed code is developed according to the changes that are provided to the development organization
- The test organization may not have been cognizant of the changes
- If the test organization is not duly informed, this will require additional time to build test cases for the added changes, or
- If the test organization is not duly informed of the requirement changes during the test phase, defects may be logged for the changed code that did not support the approved requirements.
- Dialogue occurs with the development organization about the added code that does not support the requirements that were provided to the test organization
- Dialogue occurs with the business team to determine if the new code shall be included in the current release.
- If it is determined the code must remain in the current release then:
- The change request form must be drafted, reviewed, and approved
- The requirements documentation must be updated

- The traceability matrix must be updated for the change
- The code must be reviewed to ensure it meet the approved change request
- Test case(s) must be created, reviewed, and approved for the change
- The defect is changed to a fixed state
- The defect is closed by the tester that logged the defect.

All of this extra work can be avoided if the change to requirements is followed according to the change management process. The cost to the company would be the cost of production, which consists of the cost of producing a product. And producing a product right the first time (RFT) includes the cost of labor, material, and the equipment needed to produce the product. When the product contains defects, then the cost of quality (COQ) is impacted. The cost of quality includes the cost of producing the product, labor, and maintenance. Additionally, it includes the cost of repair when defects are discovered.

Another scenario is when the tester finds some changes in the application that did not support the requirements because the developer thought it would be nice to have in the application. The tester logs a defect against the "nice to have." Unfortunately, this does not follow the change management process either. In this scenario, dialogue occurs with the development organization about the added code that does not support the requirements that were provided. Discussions are held with the business team to determine if the new code shall be included in the current release. If it is determined the code shall remain in the current release then:

- The change request form must be drafted, reviewed, and approved
- The requirements documentation must be updated, reviewed, and approved for the change
- The traceability matrix must be updated for the change
- Test case(s) must be created, reviewed, and approved for the change
- The defect is changed to a fixed state, and
- The defect is closed by the tester who logged the defect.

When the aforementioned scenarios occur, this impacts the test organization in meeting the project deadline. One thing management does not wish to do is to extend the project deadline. Additional resources may need to be added in order to meet the deadline; however, this is not always the best solution, as they are not up to speed with the business and technical needs. It is not uncommon for the test organization to be requested to work extended hours. I cannot begin to tell you how many times I have been asked to work nights and weekends over the years.

When management has requested my team to work extended hours, I have tried to be there with them, as it is not fair for management to request something that they themselves are not willing to do. Your staff is giving up their personal lives to meet the objectives of the project. Anything a manager can do to alleviate the pain is worthwhile in lifting up the spirit of the tester. It certainly helps the motivation of the team when you are there, and to bring in some food or beverages to support them.

Change can be a negative influence on individuals, and whether that is working extended hours or weekends, the pain is felt. The types of pain that can be encountered on projects when change is not managed properly are:

- Unit testing is reduced due to timelines
- More defects encountered due to last minute changes
- Increased rework
- Failure to meet timeline milestones
- Increased costs of production
- Defect removal efficiency increases rather than decreases

In my experience, it appears that there needs to be an education process with the business side of the company for them to understand the impact of adding, deleting, and changing requirements after the requirements have been approved and the project has moved onto another phase of the development cycle. Frequently, the business stakeholders have no idea on how their demands are impacting the IT organization. On my last project, the business stakeholder still had many change requests after the requirements were approved. The business stakeholder delivered a tirade on the business analysts and was relentless in making demands on the additional changes. The changes could very easily have been incorporated in a minor release after the planned release date. Not only did these demands extend the requirements phase, but they impacted other milestones for the project.

Even worse is when the project is already in the test phase and the business stakeholder is requesting unnecessary demands that must be in the current release. This results in the test organization having to respond to these demands in authoring of additional test cases. Trust me; this is very frustrating to the test organization, and most likely they are already under the gun from previously approved changes. Now they have to scurry to handle the additional changes by drafting additional test cases. This activity may require some additional time and research. The constant changes impact the quality of the software that is being tested for release, as the testers are already worn out, when extra hours are being requested.

Regression testing is most definitely impacted, as with the changes additional regression tests have to be identified and scheduled, and sometimes that is difficult to achieve due to the amount of time allocated for regression test. If manual testing is being executed, there is the probability that all the test cases will not be executed, and this opens up more defects that may be found by the customer, who will question the products that are being delivered by the test organization. Unfortunately, the business stakeholder is not held accountable for the changes that are impacting the product's quality when scope creep occurs. Obviously, there is a need for more training and coaching of the business stakeholder to understand the impact to the business with unreasonable change requests.

It is very common and a very successful method for an IT organization to use an automated tool's test repository for the storage of its test cases. In reviewing test cases that were stored in a test repository, it appeared that a number of the test cases did not support all the change requests that were occurring for a project. In those circumstances the IT organizations did not have an automated tool that would have provided traceability to all work products, such as requirements, design specifications, use cases, code, test cases, and change requests. Traceability ensures that all affected artifacts are modified and test cases are written or modified to support the change request.

Unfortunately, change management is sometimes viewed as a lot of red tape; however, failing to follow the processes and procedures of change management can result in significant disruptions for a project. Therefore, project management, the business stakeholders, and QA and IT departments need to work together collectively to develop a plan for handling of changes. When the plan is in place, work the plan.

Change is something can be positive as long as it is managed properly, and Lyndon Johnson so aptly stated, "We must change to master change."

COMMANDMENT SEVEN:

Test Execution Must Be Managed

Test execution is a time when all the planning for test becomes a reality for the test organization. All the test planning and test procedures have been identified and documented, and now the test organization is ready to begin test execution of the software application. I have heard testers say "the fun begins" because they are now ready to perform what they were hired to do and that is to measure the quality of the application.

The test manager must have the ability to deal with complex situations prior to testing and during test execution by coordinating activities, negotiating with other managers, and having the strength of character to state the application is a "Go" or "No go." And when it is a no go, the test manager must have all of his or her ducks in a row by providing all the results that support the no go decision.

The test manager is accountable for all the planning and test execution activities, and those activities may include:

- The setup of the test environment
- The setup of the data that will be needed for test
- The planning for thc type of tests that will be executed during system test, such as:
- Dry Run, Sanity, or Smoke test
- Functional test
- Structural test
- Regression test
- Performance test
- "Ility"
- Web specific testing, i.e., html syntax checking broken links, interfaces, operating systems, browsers, graphics loading, etc.), and
- Interface validation of outbound and inbound data by the test team and the user group includes:
- Scheduling when these distributions will occur
- Determining who will have responsibility
- Coordinating activities with the recipients of the data for validation

- Determining the number of test cycles that will be executed for the test
- Automated test
- Manual test
- Test tools
- Resources–Employees and contractors (this may include offshore resources)
- Communication process for the builds into the test environment
- Execution and defect metrics for the go/no go meeting to install the software in production or readiness to send to the client

For the resources whose designated role and responsibility is to test the software application, the focus is totally on the product that will be installed in the test environment. The tester's role is to measure the quality of the application. When I was performing testing, for me, this was a time of excitement, as this was a time to iron out all the defects that may be present in the application, and to ensure the product that was being delivered to the customer was a quality product.

Automated Testing or Manual Testing

The test manager will have to decide if an automated test tool will be used for the system test phase and for what type of test. Automated testing has changed over the years from a record and playback tool to keyword-driven tests that are based on objects. Certainly, automated testing has become more effective instead of only performing record and playback of the software test.

Those who feel automated testing was the silver bullet to expedite the building of test cases are somewhat mistaken. Not all tests can be automated, and the creating of automated test scripts can take longer than the development of manual test cases. However, in retrospect, the execution of automated test scripts is much more expeditious and reliable than the execution of manual test scripts.

There is a significant cost for automated tools, and there must be a budget to handle the purchase of this type of capital expenditure and the maintenance charges for the tools. Additionally, there are the training needs to use the tools and the budget that would allow for training. If your business is considering purchasing an automated tool, this activity must begin way before the project begins.

I was a QA Manager for a dotcom company, and I proposed the concept of purchasing an automated tool for the business. It took months of planning for this activity to move forward. For this activity alone, it took months of meeting with the various tool vendors and proof of concepts to occur before a purchase could be initiated. Once the cost of purchasing and training costs had been approved, then training could move forward for the QA and Development teams.

Today, there are open source (freeware) tools on the Web, and I have known of organizations that have downloaded them and have used them for unit testing and functional testing. Downloading of open source tools will depend on the policies established by the organization, as some companies do not allow any open source tools to be downloaded.

There are some very effective imaging tools for testing Web sites, as one of the areas of pain is creating and executing all the tests for the browsers and operating systems that may be used by the user. This can be a very time-consuming activity, and the use of imaging tools may reduce the testing effort and the amount of hardware to meet all the configuration needs. However, when a tool is selected, the test manager must have the staff that has the skills to use the test tool of choice.

Manual testing can take a significant amount of time to execute a test case (depending on the number of steps in the test case), and a lot of steps in a test case can be very time intensive to execute one test case. One of the issues with the execution of manual tests is the tester may get bored or complacent during testing and not be as cognizant of comparing the expected results to the actual results of the test. The size of a test case will need to be evaluated as to the amount of time it will take to execute the specified test. Manual testing may require additional test engineers to be brought on board to meet project deadlines. The testers that are added to the project may not be as acquainted with the requirements and understandings of the application and miss defects in their testing efforts.

One of my most successful consulting projects was when I worked as the business and quality assurance analyst for a software vendor that developed mainframe software products for insurance companies. The project was the addition of a commercial lines insurance product to the current lines of products the business offered, as well as to future clients. We had four months for the project, and this included the authoring of requirements, meeting with the client, performing a walkthrough of the requirements document, delivering design documentation for signoff, receiving signoffs by the client, construction of the application, and testing the software. The amount of allocated time for the system test was two weeks. We completed testing on time, and all severity levels of reported defects were resolved by the development team; the project deadlines were met. Five years later I had returned to the software vendor and the manager that I had worked for on the project and was asked if I was aware of what I had done when I worked for him before on the product mentioned above. Of course, I was not sure what direction he was going, and he then informed me that in the history of the company, the delivering of a defect-free line of business had never occurred and nor had it occurred subsequently.

Testing Must Be Scheduled

Before tests can begin, the test execution phase must be scheduled in a project plan. All the tasks associated with testing must be planned for all test types with dates for manual and automated test execution's start and end dates. In my experience, I have found that the amount of time for test planning and test execution is underestimated. So often the timeframe for testing is not calculated based on the size and scope of the project but from agreements made with the customer before sizing is determined. Unfortunately, the test execution planned dates are most often backed into based on the amount of time that has been allocated to requirements and development phases of the project, and then the QA organization has to fit their planned dates within the amount of time that is left before the delivery date.

It is also not uncommon for the timeline to be reduced because development has taken longer than anticipated to develop the logic, and then the QA organization is asked to reduce their amount of time for testing. The QA manager must then rethink the test execution plan and redraft an updated plan for test execution's end date.

The test schedule will be documented with the manual tests that are planned to be executed during the test execution phase, and the automated tests are scheduled in the automated tool that will be used for test execution. Again, the problem is when tests are scheduled for a specific day and time, and there is the probability that the schedule can go awry due to defects that are found during the test process. A word to the wise is to be prepared to address changes in schedule before it happens, as the test manager will need to prepare a mitigation plan to address this issue.

I was on a project where my responsibility was to do the planning for an acceptance test. The system test organization very proudly stated that they had every test case scheduled down to the day, hour, and minute. The problem with this plan: it can very easily go awry. On the very first day there were problems encountered when the first test failed, and the system test team was not able to continue testing for a number of days because of the issues that were encountered with the Web site.

Once this type of problem occurs, this throws the schedule completely off. Then the laborious task of updating the schedule becomes a very time-intensive task; therefore, in planning a minute-by-minute test schedule management must consider the likelihood of defects occurring and allow for that in the test plan schedule.

Testing Must Be Managed

The test manager must be cognizant of the status of test case development to ensure all the planned tests will be ready for test execution. In the situation where offshore teams are used for the authoring of test cases, it is wise to have daily status meetings that will address the state of test case development for manual and automated tests. For this daily meeting, an agenda should be prepared and distributed to all respective parties.

If the application contains a number of components, such as customer care, billing, shipping, etc., the test case readiness for each of the components should be reported for each of the components to ensure all tests are being created. Any issues or risks should be addressed in this meeting for management to immediately follow up on.

Action items are logged, and a resource is assigned to handle the action item. In follow-up meetings, the action items are reviewed to determine the state of resolution.

The test manager provides the support to the QA team in managing of the activities for test execution and, if necessary, does some hands-on activities. I am not a strong proponent of a test manager also serving as a tester; however, this is a very common practice today. When businesses believe the test manager is to also perform testing with the team on a regular basis, this requirement changes the dynamics of the role of the test manager and managing of the testing staff, as now the manager is performing at the levels of the testers. Any problems with a tester that would require a reprimand can become more of a difficult situation for the manager.

The test manager must be able to provide the soft skills in negotiating, serving as a facilitator in meetings, managing the test schedule and the test project plan, managing and overseeing test execution, managing the testing staff, attending numerous meetings and informing management of project issues, risks, metrics, and project tasks when:

- The test execution phase timeline may be reduced due to the amount of time it has taken development to construct the software
- The testers have to determine which tests will not be executed due to the amount of time that is allocated for testing

- The testers are requested to work longer hours
- Not all the tests have been executed and delivery is being requested
- There are outstanding defects, and project management would like to deliver with unresolved defects
- The test engineers have reported the product is not ready for delivery, and
- The QA Manager has to have the strength of character to say, "No go."

During test execution of the test phase, the test manager and test leads address the status of test execution with management. The test execution results are discussed and comparisons are reviewed to the planned schedule for:

- The number of executed test cases
- The number of test cases that were not run
- The number of test cases that were executed and were blocked due to a defect that was reported affecting a dependent test case
- The number of test cases that failed
- The number of test cases that were executed successfully

This discussion is not necessarily focused on defects; however, if there are defects that are affecting the failure of other test cases being executed, then this needs to be addressed. In other words, if there are defects of critical and high severity levels that are preventing workarounds, these failures need to be addressed to determine when they can be resolved and who will need to take ownership of the defect in order for testing to resume in a productive manner. Remember, the purpose of the above discussion is to determine the status of test execution and communicate those results to the leadership and other departments. This meeting also provides a benchmark as to how far the test organization has proceeded in the execution of test cases, and this will assist in determining the readiness of the application for delivery.

On a consulting opportunity, it was discovered during the test execution daily meeting that the offshore team was not executing as many tests as had been executed previously. In the discussion, it was determined that an offshore executive had reduced the number of testers without the knowledge of the company that had secured them for offshore testing because the project's timeline was expanded beyond the planned completion date. Of course, this issue had to be escalated to higher levels of management.

This issue could have been avoided by requesting an organizational chart prior to the start of the project with a list of all resources that are part of the overall test effort. Any staff increases or reductions are disseminated to management. The request for proposal should have included a request for the vendor to include an organizational chart with a list of all offshore resources that are part of the overall test effort and to include updates periodically to ensure all resources are being accounted for until the project is completed. This, I realize, is a question of trust, but sometimes it happens as addressed above and will have to be managed carefully.

I recall a number of years ago when I was managing the compliance validation team for the Y2K effort of an airline. I had encountered an issue, and one of the managers came to me and said, "Work the problem, work the problem, and you will come to a solution." When I have encountered problems or issues during the test process, that manager's word of wisdom comes to mind, and if you follow the process of issue resolution, it works. What is important is the manager cannot avoid issues, as this can have a negative impact to the project.

As a manager you may need to find out who needs to be involved in resolving the problem. The issue may be of a nature that the issue will need to be escalated, as in the situation with the offshore team's management reducing the number of testers during test execution. In this circumstance the issue needed to be escalated immediately. Your next level of management needs to be apprised of the issue and your intent to escalate, as you will need their support.

Neal Whitten writes:

> When two parties cannot agree on an important issue, escalations should be set up quickly. If the issue is truly important, a decision must be reached so that everyone can get on with their business. As a guideline, escalations should be initiated *within two working days* of being identified. Immediately,

the next levels of project leadership for whoever gets involved in your organization must reserve time on their calendars for the escalation meeting. Although an advance notice of two working days might not always be enough time to get an appointment on some calendars, management must support the expedient resolution of escalated issues.

Some guidelines to help address the important area of escalations are:

- Escalate only after a sincere attempt has been made to resolve the issue
- The dissenter is responsible for escalating the issue
- Initiate an escalation within two working days of knowing the problem is irresolvable at the current level
- Escalate the problem, not the person
- Always inform your management prior to initiating an escalation; you will need their support
- Always inform involved parties before beginning the escalation; the goal is no surprises
- When an escalation is being pursued, do not stop working the plan-of-record[30]

Manage Regression Testing

As stated earlier, the definition for regression test is selective retesting of a system or component to verify that modifications have not caused unintended defects and that the system or component still complies with its specified requirements. Regression testing becomes a necessity whenever the code is changed, whether that is due to requirements changes or when a defect has been fixed. If this is true, when testing is being performed a new build is moved into the test environment in which a regression test is to be performed. This can be very time consuming when a project is on a tight timeline, and a regression test may take a great deal of time to re-execute all the test cases manually.

In planning for a regression test, it would be a prudent move to allow time in the project plan for regression test preparation, regression test execution, and the probability for creating additional test scripts for changes (fixes to defects or last minute requirements changes that are approved for the current release).

If an automated tool will be used for regression testing, consider the following questions:

- Are there test scripts that can be reused from previous releases?
- Are you going to use the automated test scripts for the current release?
- Are you going to use the automated test scripts for subsequent releases only?

- How many of the test scripts will need to be updated based on the modifications to the application?
- Have you determined which tests can be executed using an automated tool, and which tests will be executed manually?
- Are there test engineers that have the skills and knowledge to create, execute and analyze the test results?
- Will there be a need for additional hardware?
- Will there be a need for another test environment to execute the regression test scripts?

The test manager will need to work closely with her team to plan the regression as it is not a time to have resources on the team that have a partial knowledge of authoring test scripts and will need training with the automated tool of choice.

It is not uncommon that the scripting of tests with an automated tool may take longer than preparing test cases manually. I cannot say this enough: ensure a proper amount of time is allocated for test preparation when an automated tool will be included in the test. In retrospect, the execution of the automated test scripts for regression test saves a tremendous amount of time rather than having a group of testers executing them manually.

Regression testing may require more than one regression test, and when there are many iterations of code being moved into the test environment, the number of regression tests will need to be evaluated and determined. How many regression tests are needed for an application? If we look at the pure definition of a regression test, it appears that every time the code is changed, a regression test will need to be performed. The test manager or QA lead will have to determine the number of regressions that can be performed based on the timeline. If the amount of time that has been allocated for test is reduced, then the probability of defects not being discovered because some test may not be executed, which increases the risk of the customer receiving software that contains defects and may impact their business negatively.

If the timeline for test is reduced, the test manager needs to address the concept of risk-based testing for regression test. A meeting will need to be called with the test team to identify which test cases or test scripts will not be executed and what will be the impact to the business if the test cases are not executed. In this circumstance, the test manager and/or the project manager will need to convey to the business the reduction of the timeline, the degree of risk that may impact the business when regression testing is reduced, the functional tests that will not be executed, and convey the probability of risk to the business.

If the developer states, "I only changed one line of code," will a full regression test need to be performed? To say that a full regression test should occur depends on a lot of information. If the logic is object-oriented code and the defective object is encapsulated, then the tester will only need to retest that portion of the code where the defect was discovered. If the object is not encapsulated, then this will require some knowledge about object inheritance and to determine what other testing will need to occur to ensure the defect has been resolved. The tester will need to determine whether the one line of code change will have an impact on other areas of the application. It will also require the tester to repeat the test as exactly as originally initiated when the defect was discovered.

Let's look at a couple of scenarios where defects are discovered in test cycles:

Scenario Number One

Test Case Number	Cycle I	Defect Resolution	Cycle 2	Defect Resolution	Cycle 3
1	Passed		No Retest		No Retest
2	Failed	Fixed	Re-Tested & Failed	Fixed	Re-Tested & Passed
3	Passed		No Retest		No Retest

Figure 7.1 Scenario Number One - Defects in Regression Test Cycles

In *Figure 7.1* we observe that all the test cases passed during the regression test cycles; however, is that really true? Not every test was re-executed during regression. There is a probability that another defect was introduced after the defects of test cases numbers 2 and 3 were resolved, as there may have been a dependency to other parts of the code that a defect could be residing.

Scenario Number Two

Test Case Number	Cycle I	Defect Resolution	Cycle 2	Defect Resolution	Cycle 3
1	Passed		Passed		Passed
2	Failed	Fixed	Re-Tested & Failed	Fixed	Passed
3	Passed		Failed	Fixed	Re-Tested & Passed

Figure 7.2 Scenario Number Two - Defects in Regression Test Cycles

Figure 7.2 reveals that all the tests cases were executed for the three cycles. It appears that with the execution of all the test cases for all the three cycles, there were defects in residence that had not been discovered in some of the previous cycles. Scenario number two outlines how a defect encountered for a test case that had previously passed in a previous cycle may cause a defect in subsequent cycles. This may occur when there is a dependency to other logic in the application.

To assume a defect is fixed may increase the probability of risk to the customer. Therefore, before the application is handed over to the customer, management and appropriate staff need to review the test results and determine if the risks that are identified would impact the go/no go decision in releasing the application.

Risk–Based Testing

In order for risk-based testing to be considered, an early risk analysis will have to been formulated. The business organization will have to have documented the priority to the business when the requirements were drafted.

When timelines are a factor, management needs to consider the benefits of risk-based testing. Some of the benefits of risk-based testing are:

- Testing activities are focused on the requirements with the greatest amount of impact to the business
- Provides a method for the business, development, and test teams to align on assumptions of risk that will drive the test approach and planning process
- Increases test effectiveness while potentially resulting in fewer tests
- Potentially reduces cost while maintaining or increasing test effectiveness
- Enables management to accurately assess the impact of effort, and the impact of risk to the project

If it is decided that risk-based testing is the approach that will be taken and a priority has not been assigned to each requirement, this may require some additional time to update the priority status of each of the requirements; however, risk-based testing would reduce some of the test cases that would be executed. Then management would need to come to a conclusion that risk-based testing is the approach they are willing to move forward with for the test execution phase.

Performance Testing

Of all the types of tests, this has a higher degree of complexity. The activity requires the knowledge and understanding of what the various performance tests will do—how the various types of tests can provide the needed information to "tweak" the servers or the hardware platforms to their maximum potential.

If the test manager has experience in testing but has not performed automated testing, my recommendation would be to take a class in the tool of choice to be able to understand what needs to be considered for performance test, to ask the right questions, and to be able to understand what the testers are finding when the performance test is executed. The test findings may indicate the need for another server, and this information must be conveyed to other levels of management. The test manager will need to have on board a performance tester with the testing skills that can evaluate the performance test results and then report those results to management in a manner that management and other recipients can understand.

The test manager will have to coordinate with other departments, such as operations, network services, and database administrators for them to be included when a performance test will be executed, as they will be needed to respond to the test results.

This type of testing requires performance test engineers with a significant amount of experience in performing testing. What I have found in my years of testing and managing test teams is that the performance testers must have the programming background to build the tests correctly. Oh, yes, the tools have become more user-friendly; however, they still need those programming skills to be able to write some specific code within the tool to perform tests that meet the needs of the business. Usually, the performance tester will be working very closely with the operations/network resources, as they will be the individuals that will have to tweak the servers and/or hardware platforms.

Performance tests will have to be managed and controlled when these tests are being executed against the test environment, as it can have a negative impact on other types of tests. Performance testing may have to occur at times that will not affect the other types of tests, or a separate environment may need to be created for performance testing.

COMMANDMENT EIGHT:

Shall Not Run Out of Data during Test Execution and the Data Must Be Validated Prior to Test Execution

Data needs for testing is always a concern for the test phase. In the planning process for test execution, discussions shall occur with the database administrators to plan the data that will be needed for test. Ideally, the test environment shall contain a copy of the production data; however, when that is not possible, a number of considerations must take place.

The following are some examples of questions that will need to be addressed in preparation for system test:

- Has the data been cleansed and validated when a data migration has occurred?
- How much data will be needed for the test?

- Are the testers going to be using a full copy of production data or do they need a full copy of data for system test? If not,
- Will a representative sample of data be adequate for testing?
- How will the data be created when it is a new application and there is no data?
- Whose responsibility will it be to create the data?
- Who will have the responsibility of verifying the data?

If this is a new software application, the data may have to be created, and this can be a time-consuming effort. I was working on a project for the creation of a new product line for a commercial lines rating application. The data did not exist, as it was a new product line. The data had to be a copy of what existed in the insurance company's commercial lines insurance manual for specific state territories. This had to be created, and I recall spending many hours building the data for test execution. Of course, this is not ideal, but we have to face reality: this does happen.

Data is always an issue, and being able to extract data for the system testers to work with can be a time-consuming activity. However, this is a resource that must be planned well in advance and is not to be taken lightly in order to have a successful test.

Years ago, I was working on a PeopleSoft project where the plan was to run a parallel test of the current system and compare it to the legacy system. PeopleSoft is a COTS (customized off the shelf) product that was being installed in a very large telecom company. The manager responsible for the development project planned a parallel test, and there was a concern about the size of the number of employees in their database. To reduce the risk, I recommended a small parallel test be run first to resolve most of the defects with a smaller file and then follow up with the full file test. In order to create the data needed, it was suggested a representative sample be defined for all the employees. The representative sample consisted of five thousand employee records to meet all the test conditions for test execution .

On another project that I had worked on previously as a QA lead consultant, the data conversion activities were not totally completed before the construction phase. During unit test, the developers did not have enough data to validate the outbound and inbound interfaces. The developers did their best with creating some data; however, it was not as successful for them as it could have been if the converted data had been completely validated. When the test phase began, a couple of the developers came to me to apologize for not being able to test to the degree that they felt was necessary to ensure the interfaces were functioning correctly.

Unfortunately, not all the data had been validated or cleansed (data cleansing is detecting and correcting the tables or data that are stored in the database) prior to test execution. When defects were being logged, it had to be confirmed first if it was in the code or the data. It lent itself to all kinds of problems that extended the time for defect resolution. The risk was anticipated; however, a number of the defects could have been avoided if the data conversion process would have been completed before unit testing or system test was executed.

In the above scenario, when deadlines are being determined, executive management must fully understand that processes must be in place for data validation, and the data must be validated before unit test and system test can begin. It is better to bite the bullet and allow:

- Adequate time for data validation, or
- Hire consultants that have the experience in validating converted data.

Within the plan for system test execution is planning for the data to be refreshed after a test cycle is executed. Most certainly, during test execution the data that been used in a previous test will need to be refreshed before the next test can occur.

COMMANDMENT NINE:

Reports and Metrics Shall Be Provided to Management

Whatever your role may be in the QA organization, various reports and metrics are to be provided to management to use as a mechanism in evaluating the process of test planning and test execution for a project. Your role would most likely determine the type of report or metric that will be prepared.

Most certainly there are a number of reports that many organizations have to prepare; however, I am only trying to highlight the most common reports that I have had to prepare over the years. Certainly, there are many variations of these reports, and the examples that are provided are an example that may be used. Your changes to these reports will be most welcomed by your own organization.

Status Reports

One of the reports that aids management is the weekly status report, as it provides the status of test preparation and execution to management. This report can be a simple e-mail or a template that is a corporate standard for project reporting. At most, the status report should be one page in order for the reader to quickly glean the necessary information. An example of a status report is provided in *Figure 9.1*.

<Project Name> Status Report	
Date:	
To:	<Report Recipient's Names>
From:	<Responsible Role to draft the Status Report>
Milestones:	Document milestones met, milestones not met and milestones that are due within in the next week.
Budget:	If role is responsible for the budget, then document if you are meeting, exceeding or under budget
Changes:	Document any changes that deviate from the QA Test Plan
Accomplishments:	Document project accomplishments
Comments	Succinctly document additional comments
Issues:	**Risks:**
Document issues that have occurred for the test phase of the project.	Document risks that may impact the project during the test phase of the project.

Figure 9.1 Status Report Template

Issues and risks are to be prioritized according to importance of the issue or risk. The most common prioritizations for issues are high, medium, or low. Risks are quantified according to the probability of impact to the project. The impact to the project may be defined as critical, severe, high, medium, and low. To support the process of risk avoidance, a mitigation plan and contingency plan are prepared and followed up on a regular basis to ensure the risk is prevented.

Risk and issues must be followed up and addressed in project status meetings and leadership meetings. A project I was working on for the state of New Jersey held weekly leadership meetings where all risks, issues, and action items were discussed at each meeting to ensure they were being addressed, as well as scheduled risk management meetings where the owner of the risk would address updates to a defined risk. Risks must be followed up on a regular basis as they can have a negative impact on a project.

Project Status Meetings

At the project status meetings, the management teams report the status of tasks that are due and identify any risk to meeting those objectives. This is not a time for the QA manager to be reticent about not reporting the current state. If there is any probability that a task is not going to meet the planned date, this is a time to report the status.

I recall the first time I had to report the status in one of these meetings; I was very worried that the team was not going to meet the objective of the planned date. I was certain the halls of justice would come down upon my head, but I learned that this was a time when all the participants addressed what needed to be done to meet the planned date and discussed what could be done to meet the date. Therefore, QA managers and leaders do not be fearful of reporting the status, as there is a management team to work with, and they too have an objective of meeting the project's overall objectives. However, it is recommended that your senior manager is informed prior to the status meeting of any issues that would have a negative impact on the project. This will give your manager a heads-up opportunity to support you in these meetings. Never, never put your senior manager in a position of being blindsided in a meeting.

Check Sheet/Checklist

The check sheet is a form to gather and record data in an organized manner to help determine the occurrence of an event or cause. The check sheet is also known as a checklist or tally sheet of events. An example of a check sheet is located in the chapter, "Commandment Five: Testing Shall Always Be Planned."

Project Health Report

The project health report is a variation of a status report; however, the report is a more comprehensive view of the project and provides management with information that generally a status report does not provide. The information that is addressed in this report is most likely compiled into a combined report and then disseminated to upper levels of management. The header for a project health report would contain the name of the project, the project number if a number scheme is required by the organization, the name of the project manager, the name of the senior manager, the name of business sponsor, and the name of the software quality analyst.

Status

In the status section, the author documents the current state of the project and whether the project is in a green, yellow, or red status. This status would be reported with the project manager in agreement to the status. If there is a disagreement, the author will document that the project manager does not agree to the state of the health of the project.

1. Project Plan (WBS)

Document the WBS project issues. If there are no issues, then state there are no known WBS issues.

2. Risk Management Plan

Document the project's mitigation and contingency plans. If there are no known risks, the author states there are no known project risks to report.

3. Resources

Document issues with resource needs, and if there are no known issues, then state there are no known issues to report as of this date.

4. Commitments

Document any issues with commitments, and if there are no known issues, then state there are no known issues to report as of this date.

5. Milestones/Project Plan Development

Document if there are any changes in scope and if the dates are impacted. If there are known changes, then state the known changes.

6. Project Tracking

Document the status of milestones and status of the planned dates. A graphical report is created from the project plan dates. The following represents an example of this report.

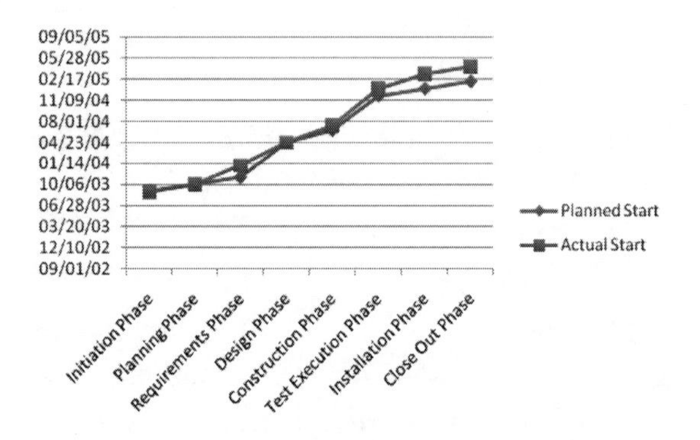

Figure 9.2 Project Milestone Status Report

7. Change Management

Document any changes in scope, and if there are no changes to report, then state there are no known issues to report.

8. Issue Management

Document any issues and any follow-ups to the issues. If there are no issues to report, then state that there are no known issues to report.

9. Root Causes

The root cause documents the root cause of the issue in salient points such as:

- The root cause is the database conversion project
- Required resources are still allocated to <Project Name>

QA Management Metrics

The QA manager's role includes the reporting of information to various levels of management. This information is gathered from client needs, status reports, other meetings, test planning, and test execution results. The type of information that should be reported is:

- QA test planning and test execution effort and variance to the effort
- Number of QA issues and the status of the issues
- Number of QA risks that have an impact to the schedule and to the business
- Number of QA action items and status

- Number of defects reported to number of test cases executed to date
- Number of defects resolved to date
- Individual project component test result report
- Tests behind schedule
- Uncorrected critical defects
- Major uncorrected defects over five days old
- Number of uncovered defects not corrected

Measure Project Size

Function points are a work product measure. They quantify data processing work outputs. Function points measure the amount of information processing functions of an application system. These measurements take into account different types of processing functions within a system, as well as the degree of influence certain system characteristics have on function complexity. When function point counts are coupled with work-effort measures (i.e., costs to build a system) and are analyzed in conjunction with project specific attributes, productivity can be measured, and the relative effectiveness of development approaches, tools, and project teams can be evaluated.

In 1979, IBM placed the function point metric in the public domain. Function points are the weighted totals of five external factors:

Factor	Number		Weight		TOTAL
Number of Inputs	10	X	5	=	50
Number of Outputs	10	X	4	=	40
Number of Inquiries	50	X	5	=	250
Number of Logical files	5	X	10	=	50
Number of Interfaces	10	X	7	=	70
Unadjusted function points					460
Complexity adjustment multiplier					1.2
Adjusted function points					552[1]

Function point counting requires specific skills to assess the number of function points for a system. It is recommended that if a company is interested in determining the size of an application, they should hire a consultant who is experienced in function point counting.

Defect Removal Efficiency

Removal efficiency is a key measurement that reports the number of defects that are found. The formula for Defect Removal Efficiency is: The Removal Efficiency = The Number of Defects Found/Number of Defects Present.

The removal efficiency formula can be evaluated at various phases in a project. For example, the removal efficiency can be determined after reviews, inspections, and tests have occurred; however, a cumulative total will provide a percentage of defects by a series of reviews, inspections, and testing of the software.

Defects found by Phases:

Software Development	
Requirements Reviews	20
Code Inspections	10
Testing	60
Defect Reported Sub-Total	90
User Phase	
Valid User Reported Defects	20
Total Number of Reported Defects =	110
Removal Efficiency	81.80%

Reports to Management

How metrics are provided to management can make the difference in how that information is digested. This information must be provided in a clear and concise manner so that the information cannot be misinterpreted. In providing metrics to management, it is recommended to present the information in a report that summarizes the results as in *Figure 9.3.* The graph displays the number of defects that were reported by category.

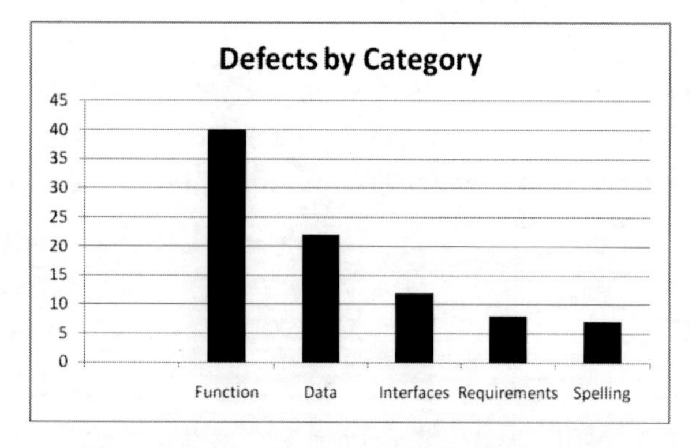

Figure 9.3 Number of Defects by Category

As we can see in the above defect report, the information can easily be interpreted without a great deal of effort. There are five defect categories that were reported, and within each category the reader can interpret the number of defects that were reported for each category.

There are test measurements that will provide the status of the testing effort and aid in the process of problem solving. These quality management tools will assist management in evaluating the process of test design and test execution.

Schedule Variation determines the capability of on-time delivery of designed tests to the number of actual tests that are available for test execution.

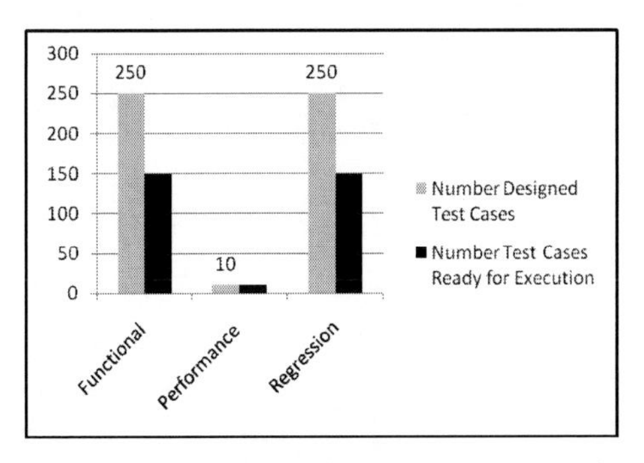

Figure 9.4- Schedule Variance–Number Designed Test Cases to Number of Actual Test Cases Ready for Test Execution

Test design productivity reports the number of test cases designed in a person day.

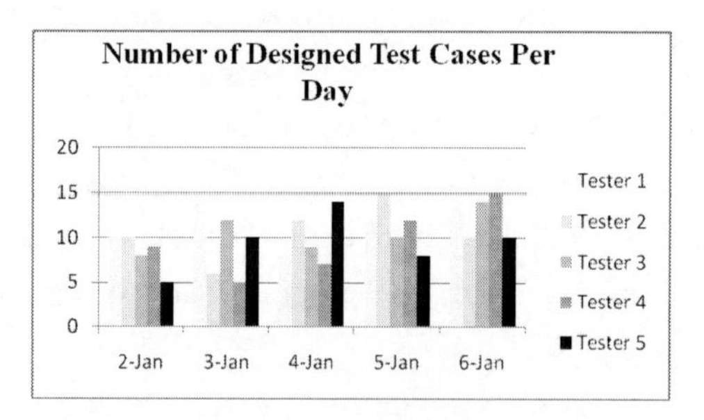

Figure 9.5 Number of Designed Test Cases Per Day by Tester

Test execution productivity identifies the number of test cases executed in a person day.

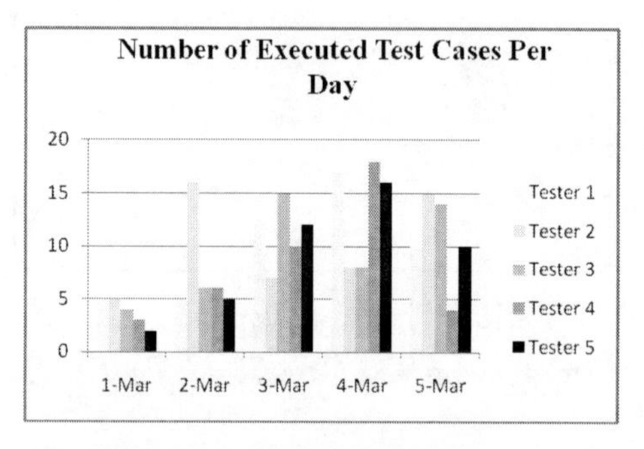

Figure 9.6 Number of Test Cases Executed Per Day by Tester

Percentage of test cases executed determines the progress of test execution. This would include manual and automated tests.

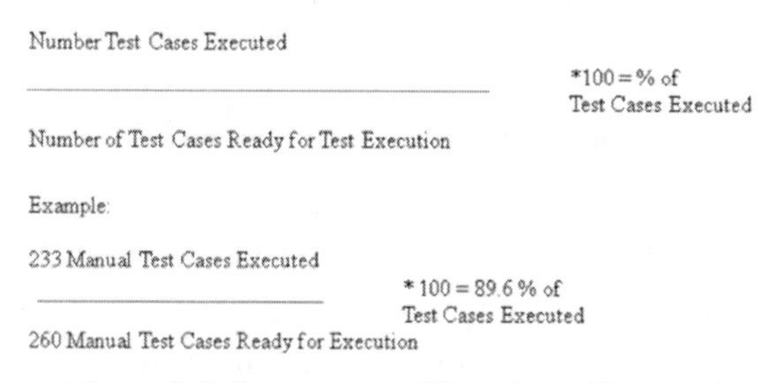

Figure 9.7 Percentage of Test Cases Executed

Percentage of test cases passed provides a test execution measurement of the software application. This measurement would include manual and automated test cases. To utilize this metric, some careful analysis needs to be performed before calculating the quality of the application. If there are one hundred test cases that trace to the requirements, and eighty of those test cases have been executed successfully, the percentage of test cases would represent 80 percent. However, the other twenty test cases that would trace to the requirements that have not passed may contain tests that would validate requirements that would have a critical or high impact to the business needs. Additionally, the analysts must review the traceability of the test cases that were not executed and the business priority of those test cases.

Problem-Solving Metrics

These tools are most definitely important tools for performing analysis; however, I must be honest with my readers: I have to wonder how much quality assurance analysts really use them. We are supposed to use these tools, but most often the metrics that can be created from the automated test tools are leveraged in providing metrics to management. However, I felt it was important for the reader to have some understanding of these very important analytical tools.

An analytical tool that aids the QA resources in problem identification is the Pareto chart. The *Pareto Chart* is designed to rank items by frequency and is commonly known as the 80–20 Rule. This analysis tool is one of QA's most important tools and is helpful in defect analysis.

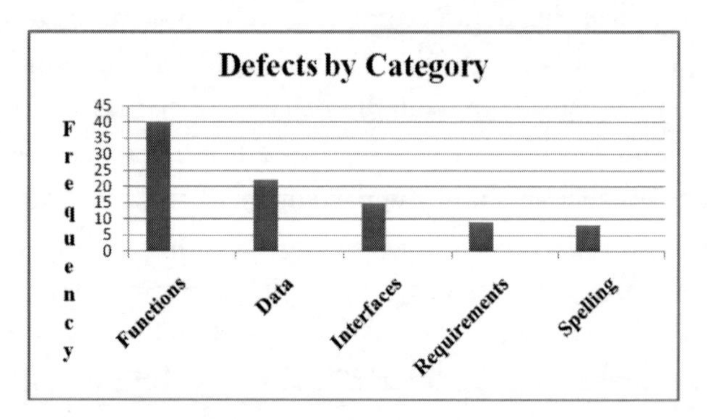

Figure 9.8 Pareto Chart

The Cause and Effect Diagram is also known as the Fishbone Diagram, as the diagram resembles a fish's bone structure. This diagram was championed by Kaoru

Ishikawa, a statistician and quality expert in Japan. This diagram aids in determining the possible causes of a problem and is created in a brainstorming session to examine the factors that may have influenced the problem.

The problem is described in a box at the head of the diagram; then a line is drawn (to represent the spine of the fish) with an arrow pointing toward the head, forming the backbone of the "fish." The direction of the arrow indicates the causes that feed into the spine that may have caused the problem described in the "head." The large bones intersect with the spine at an angle. The large bones represent the main categories (category 1, etc.) of potential causes of the problem. The arrows represent the direction of the action; the items on the larger bones are thought to cause the problem in the head. The small bones represent deeper causes of the larger bones. Each bone is a link in a Cause-and-Effect chain that leads from the deepest causes to the targeted problem.

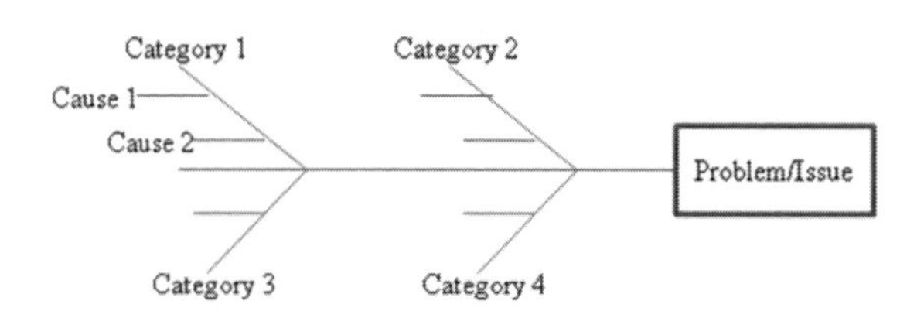

Figure 9.9 Cause and Effect Diagram or Fishbone Diagram

A tool that aids the QA resource in problem analysis is the *control chart,* also known as a *run chart.* It is a statistical technique to assess, monitor, and maintain the stability of a process. The objective of this quality tool is to monitor continuous repeatable processes and the variation from specification limits. In developing this metric, the upper control limit (UCL), the lower control limit (LC), and the variants from the standard average. The two types of variation are: common, or random, and special, or assignable cause. The bulleted list provides an explanation of variants that are measured in a Controlled Test in reporting a test run:

- If the data fluctuates within the limits, it is the result of common causes within the process (flaws inherent in the process) and can only be affected if the system is improved or changed.
- If the data falls outside of the limits, it is the result of special causes. The special causes can include bad instruction, lack of training, ineffective processes, or inadequate support systems.
- The special causes must be excluded before the control chart can be used as a monitoring tool. As an example: staff may need better instruction or training, or processes may need to be improved before the process is under control. Once the process is under control, samples can be taken at regular intervals to assure that the process does not fundamentally change.

A process is said to be out of control if one or more points falls outside the control limits.

Manual Test Measurements

Management will need to know the status of manual test preparation and the suggested bulleted list provides some measurements that would provide this information:

- Number of manual test cases per requirement in process
- Number of manual test cases per requirement completed
- Number of manual test cases per design requirement in process
- Number of manual test cases per design requirement completed
- Total number of manual test cases completed
- Total number of manual test cases not executed
- Total number of manual test cases executed
- Total number of manual test cases not completed

Automated Test Measurements

Management will need to know the status of automated test preparation, and the suggested bulleted list provides some metrics that would provide this information:

- Number of automated test scripts per requirement in process
- Number of automated test scripts per requirement completed
- Number of automated test scripts per use case in process
- Total number of automated test scripted completed
- Total number of performance test scripts in process
- Total number of performance test scripts completed

COMMANDMENT TEN:

There Shall Be Defect Management

Defect prevention is a quality effort that all organizations must incorporate as a best practice. In order for defects to be reduced, executives need to mandate a defect prevention program that includes reviews and/or inspections for all artifacts that are created throughout the software development life cycle. It is a lot less costly to resolve a defect found during the requirements phase, which is the first point of insertion, rather than in subsequent phases of the software development life cycle in whatever methodology is being employed.

For the test phase of the project, defect management includes all the activities that are involved in the reporting of software defects including:

- Finding defects
- Logging defects
- Defect analysis

- Managing the reported defects
- Creating defect metrics
- Managing the defect meetings
- Defect resolution process
- Retesting of the defect
- Managing of the changed code

There still are misnomers about quality assurance, as I have found over the years that quality assurance means testing, and to resolve this myth I would like to define the difference between quality assurance and quality control. To understand those differences is to first look at the definitions:

- *Quality assurance* is a planned and systematic set of activities necessary to provide adequate confidence that requirements are properly established and products or services conform to specified requirements.
- *Quality control* is the process by which product quality is compared with applicable standards and the action taken when nonconformance is detected.

Differentiating Quality Control and Quality Assurance

Quality Control	Quality Assurance
Relates to a product	Establishes processes
Verifies the product meets the business and design requirements	Sets up measurement programs for process evaluation
Is conducted by inspection and testing of the software to identify defects	Identifies weaknesses in processes and improves the processes
Is the responsibility of the tester	Is the responsibility of management, and is most often performed by a member of the staff
Is concerned with a specific software application	Is prevention driven
	Evaluates whether quality control is working

Figure 10.1 Differentiating Quality Control from Quality Assurance

It is the role of the software tester to validate that the software application is functioning per the business and design requirements. In order for the software tester to draft the test cases, the requirements must be written to the granularity that the requirement is testable. If the requirement contains words such as, *and or*, or, the requirement is not testable as there is more than one requirement with the inclusion of those terms. This creates an ambiguous test condition. Similarly, it is true for the expected results in a test case. In the test case steps, the expected results field shall contain one result for a flight number—not two different numbers. If the expected result contains greater than one result, then a defect should be filed against the test case step.

The expected result field must contain what the requirements would have expected to appear in the field or output data and not contain any condition statements. If the expected result field contains condition statements, this would be an incorrect test case step and should have been reported as a defective test case during the review process of the test cases.

As a QA manager for a Fortune 100 company that used only offshore resources for the testing efforts, I was reviewing test cases and their test steps that were created by one of the offshore companies when I discovered that the expected results contained condition statements. I wrote the test cases up as not acceptable and reported my findings to the offshore team. In next redraft of the test case, the same problem still existed, and I again returned the test cases as not acceptable. After the third try, I met with one of the managers of the offshore team to resolve this problem and offered to hold a training session on test case development.

This CMMI Level 5 offshore company's review process missed the condition statements and realized they needed to be more cognizant of the test steps that were being written. Test cases can be defective, and it is important the test cases must be part of the review process before the test cases are to be executed.

Admittedly, most testers enjoy the test execution phase, and to say they are only defect driven is not entirely true. They are driven that the product meets what the customer wants, and finding defects is part of the process. Good testers are worth their weight in gold, and executing tests correctly is important to them. Testers who execute tests that do not follow the process or perform what I call "funky" testing are not necessarily good testers.

I am not an advocate of what I call "funky testing." I do not believe in hitting any keys on the keyboard to cause a failure in the application. This type of testing in most cases is not repeatable. Every test that is executed must be repeatable. When a defect is encountered, the tester must be able to repeat the defect. If the defect cannot be repeated, then it is not a defect that can be reported to the development organization. When "funky testing" is allowed in system test or acceptance test, it ends up being a time-intensive task that is costly to the test organization and to the development organization.

I was on a project for a major Fortune 500 company that had hired a number of college graduates, and our test organization selected one of the grads as a tester. This individual had no prior testing experience. She was tasked with testing one of the software applications and was just hitting any keys, which caused a system failure. Even though the defect was not repeatable, the senior tester who had been mentoring the new tester stood her ground and felt the defect needed to remain on the books. The defect was a core dump, and according to the established standard, if a core dump occurred, the application could not be installed in production.

At the change control meeting, the defect was discussed, and the development organization presented their position that they had tried on numerous occasions but were not able to repeat the defect and requested permission to install the application in production. The meeting included the QA manager and senior test analyst, the network systems manager, the database administrator, the development project manager, and the customer. It was determined that the product would be installed in production; however, the defect would remain open. A year or so later, the project manager asked me if I would be willing to close the defect. I reviewed the defect and attempted, unsuccessfully, to recreate the defect. I discussed the defect with the project manager, and collectively we agreed the defect should be closed.

A lot of time and effort went into the analysis of the core dump defect that was reported based on "funky testing." My opinion is that unless a defect is repeatable, the defect shall not be reported. It would be the responsibility of the test manager to convey to the test organization that all defects that are reported must be repeatable before they are considered a valid defect.

In interviewing for various projects that I have worked on as an employee or a consultant, one of the questions addressed the area of soft skills, which include the ability to work well with members of a team and other departments. This question really is a double-edged sword for testers. I would like for you to think about this, for when the developer is constructing the code for the end product, the developer may become very protective of their effort, as this is their baby. The developer is supposed to perform unit testing for the code that was constructed. Unfortunately, it is not uncommon when construction has taken longer than planned that the application is thrown over the wall to QA with minimal unit testing, and it is up to QA to find the defects the development organization should have found in the construction phase of the project.

Even when a developer has performed unit testing, there are still defects that are encountered. Earlier, I wrote about a product that was being built for a major Fortune 500 company that was shelved after a period of time. As I had indicated earlier, there was an attitude that I, the senior QA analyst, was not welcomed as a matrixed member of the project team. The first day of testing, a significant number of defects were reported. As the senior QA analyst, I met with the lead developer to discuss some of the defects that were discovered to quickly encounter a very angry man who chose to scream at me to the point that I was in tears. To say that soft skills were being used here is obviously not true. The test team was performing its role and logging defects as they were found. As testing progressed for the first week, hundreds of defects were logged, which added to the indifference of the development team. The soft wall became a very large brick wall with the development team. It is not uncommon that a wall develops with the development organization and the QA organization because the

QA organization is measuring the quality of the code the developers have created. At first, the wall is sort of an invisible wall that is not uncommon to evolve into a brick wall. Too often they act like prima donnas. Of course, the objective is to avoid this from occurring, but it is difficult in some organizations.

Obviously, the test organization was performing their assigned responsibilities in discovering defects and reporting them as required by the standards that were established and approved by management. This became a huge management issue, as the development organization chose not to work effectively with the test organization for the specified project. In this situation, there should have been a meeting with the test organization's management team and the development management team to resolve the differences. Even though status reports were filed weekly to immediate management and to the director, the problems with the development team were not rectified or ever managed properly.

To avoid the aforementioned situation, a number of skills could have been employed if the organization had been receptive. Obviously, management skills were needed to play a more responsible role by working with the development team to be more receptive to the QA team, and soft skills, not just on the QA side of the team but also on the development side, were needed. It is commonly stated that the QA analyst must be careful in how he or she works with the developer so as not to irritate him or her. Well, this is not just a one-sided situation, and it is time for the developers to grow up and not be such babies.

As we have read, the developer became very sensitive to the number of defects that were found with the code. Since the developer was the lead developer, he felt a greater responsibility for the number of defects being reported. To avoid developer sensitivity, it is important the QA analyst develops a working relationship with the development organization for when the test execution phase of the project begins. An analogy of the development process I like to use is that this period of time may be similar to giving birth to a baby. The process of developing the code and the testing of the code can be painful; however, the end result is joy and celebration. Remember, it takes two to tango and two to make a baby, and if everyone can work well together, then success is born.

In the managing of the aforementioned project, the managing of defects was a rigorous task. Even though it was rigorous, this was made easier with the use of an automated tool. All the defect information was logged into the tool and then the defect was sent to the person who had been assigned to resolve the defect.

In assigning defects to an assignee, the assignee may not be the person who will fix the defect. In a large organization, the assignee usually has the technical experience and would have the responsibility to review the defect first before assigning the defect to another developer for defect resolution. I was working on a project where it was decided that the project management office resource would be the assignee for defects. The problem with this was the individual was tied up in meetings a greater share of the time and was not always available. Therefore, the person who is assigned to reassign the defects must be a person who is available and knowledgeable. In smaller organizations the assignee may be the developer who will fix the defect. In this case, the QA tester will then assign the defect to the developer who will be resolving the defect.

In some instances, some companies do not have a budget for an automated defect management tool; however, some sort of tool must be used for the logging and tracking of defects. Whether a person downloads an open source tool that can be located on the Web or uses an Excel spreadsheet, it is much better than using e-mails for the tracking of defects. I have to admit I was quite surprised when I was on a project for a very large corporation that the test organization was using e-mails for the handling of defects and then were tracking the defects on an Excel spreadsheet, which can be a very cumbersome task to ensure the defects status is up to date. Even though the development team was informed they were to use the defect tool to provide all the information for defect resolution, breaking the habit of using e-mails was a hurdle to overcome.

In the circumstance of not using an automated defect-tracking tool, not all the historical information may be captured in an e-mail. There may be multiple e-mail messages about the defect, and they may not be all contained in one e-mail thread. All the historical information may be there, but it is a cumbersome task to read through all the e-mails to determine all the facts about the defect, and there is a possibility that vital information was missed about the defect.

When a defect is resolved based on the historical information that was within the e-mail and in retesting of the defect, it is discovered that the defect is still not fixed, unfortunately the developer has to review all the e-mails to find out what was done to fix the defect. It would have been more expeditious to look in one source instead of multiple sources for discovering the defect resolution history.

Managing of defects is an activity that requires taking on the responsibility to ensure that all defects are resolved as quickly as possible. A method that will aid in the managing of the defects in a decisive manner is to have a daily defect meeting. The test manager sends out invitations to all the departments that would have a vested interest in the defects being reported and facilitates this meeting.

Before the meeting begins, a list of defects is distributed to all invitees, and the report is organized by severity and the date discovered. In the situation that the various organizations are not located at the same site, a planned conference call will be established. In preparation for the meeting, some preliminary analysis will need to occur before the discussion. When all the invitees are located at the same site, the meeting is generally held in a conference room; however, it is not uncommon for this to be a conference call. It is best in the planning that one hour is designated for the meeting, as it avoids long and unnecessary discussions. If there is a need to discuss a defect at greater length, a separate meeting should be scheduled with the appropriate resources, such as the tester who discovered the defect or the lead QA analyst and the developer who is assigned to fix the defect.

Reporting of defect metrics must be a standard process for a test analyst, and the type of metrics can be very simple in nature or more complex. Most certainly the following metrics shall be created and reviewed:

- Total number of defects per day
- Total number of defects by severity: critical, high, medium, and low
- Total number of defects by status (new, open, assigned, fixed, duplicate, tested, reopen, etc., and by severity)
- Total number of defects by severity and priority
- Total number of defects by trend
- Defect ageing report
- Total number of defects by short description
- Defect density report
- Total number of defective manual test cases by severity: critical, high, medium, low, and minor
- Number of manual test cases defective due to ambiguous requirements

- Number of manual test cases defective due to specification not included in the software
- Number of manual test cases defective due to design ineffectiveness
- Number of automated test cases defective by severity
- Number of automated test cases with a defect of critical, high, medium, low, and minor
- Trend defect report for a specified period of time and by status
- Defect density report

There is a sundry of defect measures that can be provided to management that will provide a quality measurement of the software application. The following lists of defect measurements that can be created for management are:

- *Defect Removal Efficiency*–The number of defects found in a test phase divided by the total number of defects found during the life of the product. (This is an indicator of process maturity and process improvement.)
- *Average Age of Uncorrected Defects*–Total number of severity categories aged according to the average number of days since the defect was reported. The calculation is to accumulate the total number of days each

defect has been waiting to be corrected, divided by the number of defects. Average days are the number of working days.

- *Defect Distribution Report*–These report modules that have an excessive defect rate, and thus the report would be an indicator of an ineffective architecture and is a candidate for rewrite. Be careful in using this report, as I have found that this is an indicator of a developer's skills, and most often management does not wish to target a developer. Before this measurement is prepared, it may be wise to discuss with management. Additionally, this report may be an indicator that a developer may have some training needs.

For the daily defect meeting, a complex report is not needed, but a simple report that identifies the defect number, a short description of the defect, the status of the defect, the priority, and when the defect was initially reported is provided to all attendees of the meeting.

For reporting to management, I recommend downloading the metrics to an Excel spreadsheet and organizing the data into a graphical report. The reports that will be distributed to management will need to be agreed upon before the report is sent and, of course, the level of management will determine the type of defect report.

At the go/no go meeting, the test manager is an attendee, as well as the development organization and other departments, to determine if the product is ready for installing the software application in the production environment or delivering to the customer. I am not an advocate of installing defects in the production environment, but to say it is not happening is a bald-faced lie. It happens all the time, and when this occurs the decision is made with discussions with the business stakeholders, project teams, and the test organization to determine the severity of the defects planned to be installed in the production environment and what the impact will be on the business if the users encounter these defects. In other words, what amount of risk is the business stakeholder willing to accept?

Also, there needs to be a plan of action for when the defects that are still outstanding will be fixed. The outstanding defects are documented in the QA test plan under the release notes section of the document and provided to the business stakeholder.

COMMANDMENT TEN + ONE: QUALITY CAN BE A SUCCESS *IF*

Quality can be successful; however, the software quality assurance organization that includes software quality assurance activities that are focused toward defect prevention and the test organization whose focus is measuring quality of an application will need the buy-in from executive management to support the quality efforts. In my experience, many companies will not provide the budget for the test organization, let alone a SQA team, or allocate the appropriate budget for a quality initiative.

This problem of a quality initiative has existed more than the past thirty years, as Capers Jones has written:

> The economic value of software quality has been poorly understood due to inadequate metrics and measurement practices. The two most common software metrics in the early days of software were "lines of code" and "cost per defect." Unfortunately both of these have serious economic flaws.

The "lines of code" metric cannot be used to measure either requirements or design defects, which collectively outnumber coding defects. It is not possible to understand the real economic value of quality if more than 50 percent of all defects are not included in the measurements. A more subtle problem with lines of code is that this metric penalizes high-level languages such as Java and Ruby and makes older low-level languages such as C and assembly language look better than they really are. For more details, refer to Capers Jones's book, *Applied Software Measurement.*

The "cost per defect" metric actually penalizes quality and tends to achieve the lowest result for the buggiest applications. This phenomenon is due to fixed costs associated with defect removal, such as the cost of writing test cases and the cost of executing test cases. Even in situations where the application has zero defects, there will still be costs for writing and executing test cases. Therefore "cost per defect" goes down as numbers of bugs go up. For more details of this problem, refer to Capers Jones's book, *Software Engineering Best Practices.*

The most effective method for measuring the economic value of quality is to analyze the total cost of ownership (TCO) for software applications. It will be discovered that applications with fewer than about 3.0 defects per function point and greater than 95 percent in defect removal efficiency will cost about 20 percent less to develop than identical projects with poor quality. Their schedules will be shorter by about 15 percent. Annual maintenance costs will be less by about 40 percent. The cumulative TCO of high-quality applications from the start of the first release through five years of maintenance and enhancement will be about 30 percent lower than identical projects with poor quality.

One final value point is very important. For large applications greater than five thousand function points in size, high quality levels will minimize the odds of failure. For poor quality, failure rates in excess of 30 percent can occur at five thousand function points. For high-quality projects, failure rates are usually less than 5 percent, and cancellations are due to business reasons rather than excessive cost and schedule overruns. The economic value of excellent quality is directly proportional to application size. The larger the software application, the more valuable quality becomes.

As of 2009, the overall cost drivers for software indicate why software has a bad reputation among CEOs and corporate executives. Our two top cost drivers are finding and fixing bugs and cancelled projects! It is no wonder that software is poorly regarded by corporate executives.

Table 5: The Top 15 U.S. Software Cost Drivers in Rank Order Circa 2009

- The cost of finding and fixing bugs
- The cost of cancelled projects
- The cost of producing paper documents and English words
- The cost of recovery from security flaws and attacks

- The cost of requirements changes during development
- The cost of programming or coding
- The cost of customer support
- The cost of meetings and communication
- The cost of project management
- The cost of application renovation
- The cost of innovation and new kinds of features
- The cost of litigation for cancelled projects
- The cost of training and learning software applications
- The cost of avoiding security flaws
- The cost of acquiring reusable components

Table 5 is a professional embarrassment. No true engineering discipline should have defect repairs and cancelled projects as the two top cost drivers. For software engineering to become a true engineering discipline, quality control will have to be much better than it was in 2009.

Table 6 shows a hypothetical rearrangement of cost drivers that should be a goal for software engineers over the next ten years. Our top cost driver should be innovation and designing new features: not bug repairs. Table 6 illustrates how costs should be apportioned circa 2019:

Table 6: The Top 15 U.S. Software Cost Drivers in Rank Order Circa 2019

- The cost of innovation and new kinds of features
- The cost of acquiring reusable components
- The cost of requirements changes during development
- The cost of programming or coding
- The cost of training and learning software applications
- The cost of avoiding security flaws
- The cost of producing paper documents and English words
- The cost of customer support
- The cost of meetings and communication
- The cost of project management
- The cost of application renovation
- The cost of litigation for cancelled projects
- The cost of finding and fixing bugs
- The cost of recovery from security flaws and attacks
- The cost of cancelled project

If software quality is improved, it should be possible to spend a much higher percentage of available funds on innovation, new features, and certified reusable materials.

Today's top cost drivers of defect repairs and cancelled projects should be at the bottom of the list of cost drivers and not at the top as they are in 2009."[31]

> It appears that executives may have the dream and espouse their dream about a quality initiative for the IT organization; however, the follow-through does not occur, as evidenced by Rick Craig, an expert instructor and consultant with SQE training, who chaired the Testing & Quality Leadership Summit at Starwest in the fall of 2009 when he wrote:

Marc Rene gave the first talk at the summit and challenged the group to respond to the question, "What is senior management's perception of testing?" The responses given by the attendees, who were mostly experienced testers and test managers, were probably not too surprising to most people in the business. Examples of the responses include:

- Bottleneck
- What is the point?
- Necessary evil
- Ad hoc
- Why so much time?
- Too slow
- Overstaffed
- Too many excuses
- Testing should find everything

- Quality gatekeeper
- Find bugs too late
- Testing less value than other disciplines

When I read this list, what seems clear to me is that senior management (and maybe developers, users, etc.) doesn't understand what to expect from the testing effort. I would like to discuss at least a couple of the perceptions listed above. Testers must do a better job of setting expectations. Many users and senior managers expect the testers to find all of the bugs. This is probably not possible and surely is not practical. If too many bugs are ultimately making their way in to production, then one way to reduce the number might be through additional testing and subsequent correction. But a better solution might be through better understanding the user's needs, creating better requirements, or producing better code from the onset.

Testing is not the quality gatekeeper. The purpose of testing is not to ensure the quality of the software but rather to measure its quality. It is true that the testers may find bugs that, if fixed, will ultimately result in a better product, but this is due not only to the testers but also to the programmers who fix the bugs found by the testers. Testing is just one facet of the quality solution. Responsibility for the quality of the product must reside in the entire team: users, requirements analysts, developers, and, yes, the testers.[32]

I certainly would not like to see companies get rid of a quality assurance organization, as this opens up for products to be more defect ridden. Prior to my joining a dot-com the company did not have a QA organization or an established QA governance. The product lines contained a number of defects that had been reported by the clients and were not happy in dealing with a defective website. An executive with the company recognized this need and created a quality vision for the company. I was brought in as the QA manager to build on that vision and to establish a QA team for testing and to be a part of the process improvement team that would support the quality vision. Once the quality initiative and the test organization were in place, the product lines became more stable and the clients had more constructive statements about the products.

It is very difficult for quality expectations to go from a bottom-up to drive-quality initiative. The quality vision and expectations must be established by executive management. An executive with a supplemental insurance company did have a vision for quality objectives and best practices. With this vision in sight, this company achieved a goal of being assessed at a CMM level III.

I would like to believe that management does care; however, executives have to buy in to a quality initiative that is not just a silver bullet to have passion about quality and put that passion into the most appropriate plan for their business.

As you see, a quality vision can happen, but are executives willing to go to the mat to make it happen?

> We are what we repeatedly do. Excellence then, is not an act, but a habit.
>
> Aristotle

INDEX

ENDNOTES

1 *Certified Quality Analyst (CQA) Examination Study Guide, Appendix B Quality Vocabulary,* p. B-1, Version 3 Copyright ©1998, Quality Assurance Institute ®, Orlando, Florida.

2 *Little Book of Testing, Volume 1, Overview and Best Practices,* p.4, Software Program Managers Network, Managed by American Systems.

3 Ibid, p.4–5.

4 http://www.uoregon.edu/-ftepfer/SchlFacilities/Tire-SwingTable.html

5 *IEEE Standard Glossary of Software Engineering Terminology,* IEEE Std 610.12–1990(Revision and redesignation of IEE Std 729–1983), p. 92, Copyright © 1990 by The Institute of Electrical and Electronics Engineers, Inc., New York, NY.

6 "Barry Boehn, "A Spiral Model of Software of Software Development and Enhancement," "Computer," IEEE, 21(5): p. 64, May 1988.

7 Ibid, p.65

8 Wikipedia Contributors, "Agile Software Development." Wikipedia, The Free Encyclopedia, 15 December 2009 11:40 UTC.

9 Ibid, p. 78.

10 Ibid, p. 57.

11 Ibid, p. 8.

12 Computer Aid, Inc., An IT Metrics and Productivity Journal and Special Edition, 'IFocus on Capers Jones, Chief Scientist Emeritus, SPR, A CAI State of the Practice Interview," p. 3, July 2005.

13 *IEEE Standard Glossary of Software Engineering Terminology,* IEEE Std 610.12–1990(Revision and redesignation of IEE Std 729–1983), p. 84, Copyright © 1990 by The Institute of Electrical and Electronics Engineers, Inc., New York, NY.

14 Ibid, p. 41.

15 Ibid, p. 64

16 Ibid, p. 64.

17 Ibid, p. 57.

18 Ibid, p. 5.

19 Ibid, p. 12.

20 Ibid, p. 47.

21 Ibid, p. 62.

22 Ibid, p. 63.

23 Ibid, p. 84.

24 Ibid, p. 2.

25 *Certified Software Quality Analyst (CSQA) Examination Study Guide, Quantifying Risk,* Version 3Copyright ©1998, Quality Assurance Institute ®, Orlando, Florida, pp. 4–13–2.

26 "Little Book of Configuration Management, Volume I, Overview and Best Practices." p. 3, Software Program Managers Network, Managed by American Systems.

27 *IEEE Standard Glossary of Software Engineering Terminology*, IEEE Std 610.12–1990(Revision and redesignation of IEEE Std 729–1983), p. 16, Copyright © 1990 by The Institute of Electrical and Electronics Engineers, Inc., New York, NY.

28 Ibid, p. 79.

29 Whitten, Neal, "Managing Software Development Projects Formula for Success." p. 87, John Wiley & Sons, Inc., 1995

30 Ibid, p. 169.

31 Capers, Jones, "Software Quality and Software Economics." p. 6–10, Capers Jones & Associates LLC, October 2009

32 Rick Craig, Consultant, "What Management Thinks About Testing." p. 100, SQE Training, October 2009, Orange Park, FL